How Numbers Work

How Numbers Work

Discover the strange and beautiful world of mathematics

NEW SCIENTIST

First published in Great Britain in 2018 by John Murray Learning
First published in the US in 2018 by Nicholas Brealey Publishing
John Murray Learning and Nicholas Brealey Publishing are companies of Hachette UK
Copyright © *New Scientist* 2018
The right of *New Scientist* to be identified as the Author of the Work has been asserted by it
in accordance with the Copyright, Designs and Patents Act 1988.
Database right Hodder & Stoughton (makers)

A catalogue record for this book is available from the British Library and the
Library of Congress.
UK ISBN 978 1 47 362974 5 / eISBN 978 14 7362975 2
US ISBN 978 1 47 367035 8 / eISBN 978 1 473367036 5
1

The publisher has used its best endeavours to ensure that any website addresses referred to
in this book are correct and active at the time of going to press. However, the publisher and
the author have no responsibility for the websites and can make no guarantee that a site
will remain live or that the content will remain relevant, decent or appropriate.
The publisher has made every effort to mark as such all words which it believes to be
trademarks. The publisher should also like to make it clear that the presence of a word in
the book, whether marked or unmarked, in no way affects its legal status as a trademark.
Every reasonable effort has been made by the publisher to trace the copyright holders
of material in this book. Any errors or omissions should be notified in writing to the
publisher, who will endeavour to rectify the situation for any reprints and future editions.
Cover image © Shutterstock.com
Typeset by Cenveo® Publisher Services.
Printed and bound in Great Britain by CPI Group (UK) Ltd, Croydon, CR0 4YY.
Hachette UK's policy is to use papers that are natural, renewable and recyclable products and
made from wood grown in sustainable forests. The logging and manufacturing processes are
expected to conform to the environmental regulations of the country of origin.

John Murray Learning
Carmelite House
50 Victoria Embankment
London EC4Y 0DZ
www.hodder.co.uk

Nicholas Brealey Publishing
Hachette Book Group
53 State Street
Boston MA 02109
www.nicholasbrealey.com

Also available
as an ebook

Contents

Series introduction

New Scientist's Instant Expert books shine light on the subjects that we all wish we knew more about: topics that challenge, engage enquiring minds and open up a deeper understanding of the world around us. *Instant Expert* books are definitive and accessible entry points for curious readers who want to know how things work and why. Look out for the other titles in the series:

The End of Money
How Your Brain Works
The Quantum World
Where the Universe Came From
How Evolution Explains Everything about Life
Why the Universe Exists
Your Conscious Mind
Machines That Think

Scheduled for publication in 2018:

Human Origins
A Journey through the Universe
This is Planet Earth

Contributors

Editor: Richard Webb, Chief Features Editor at *New Scientist*

Instant Expert Series Editor: Alison George

Instant Expert Editor: Jeremy Webb

Guest contributors

Richard Elwes wrote parts of the chapter on infinity and 'The algorithm that runs the world' in Chapter 8. He is a writer, teacher and researcher in mathematics, and a visiting fellow at the University of Leeds, UK. His latest book is *Chaotic Fishponds and Mirror Universes* (2013).

Vicky Neale wrote about the twin primes conjecture in Chapter 5. She is the Whitehead Lecturer at the Mathematical Institute and Balliol College at the University of Oxford, UK, and author of *Closing the Gap: The Quest to Understand Prime Numbers* (2017).

Regina Nuzzo wrote the section on frequentist and Bayesian statistics in Chapter 6. She is a writer, statistician and professor at Gallaudet University in Washington, DC.

Ian Stewart wrote the sections on the empty set in Chapter 2 and the maths of elections in Chapter 8, as well as the conclusion on what makes maths special. He is Emeritus Professor at the University of Warwick, UK, and the author of numerous books on mathematics, the latest of which is *Calculating the Cosmos* (2017).

Thanks also to the following writers and editors:

Gilead Amit, Anil Ananthaswamy, Jacob Aron, Michael Brooks, Matthew Chalmers, Catherine de Lange, Marianne Freiberger, Amanda Gefter, Lisa Grossman, Erica Klarreich, Dana Mackenzie, Stephen Ornes, Timothy Revell, Bruce Schechter, Rachel Thomas, Helen Thomson.

Introduction

In 2014 the Iranian Maryam Mirzakhani became the first woman to win the highest honour of mathematics, the Fields Medal. To her, mathematics often felt like 'being lost in a jungle and trying to use all the knowledge that you can gather to come up with some new tricks'. 'With some luck', she added, 'you might find a way out.'

Mirzakhani, who died in July 2017 at the age of 40, ventured deeper into the mathematical jungle than most. This *New Scientist Instant Expert* book is for those wandering on the periphery looking for a way in.

Willingly or unwillingly, most of us have gleaned some idea of what the mathematical terrain looks like. There are symbols, equations and geometrical shapes. There are problems with right answers, truths that are seemingly universal, and proofs that are logically watertight. Above all, there are numbers.

But how does it all hang together? What makes numbers and mathematics special – and some numbers and bits of mathematics more special than others? This is too broad a subject to hope to give a comprehensive overview but, by drawing on the thoughts of leading researchers and the very best of *New Scientist*, we hope to build up a picture.

After a brief introduction to the nature and scope of mathematics itself, we start where it all started: with the fascinating properties of numbers. We look at zero and infinity, the prime numbers, and at inescapable oddball numbers such as the 'transcendentals' e and π and the imaginary unit i. Via a brief diversion through the problems of probability and statistics, we arrive at the cutting edge of modern mathematical methods and examples of how they are applied in some unexpected areas of our lives, before considering the deepest problem of all: how exactly does mathematics relate to reality?

For many on the outside, the wonder of mathematics lies in the way it seems to be a universal language that helps us better understand the world. Many practitioners would agree, but they add that its beauty lies in how, from simple beginnings and using only the tools of purest abstract logic, you can create worlds that seem to transcend our own.

Mirzakhani studied the geometry of a thing called moduli space, which can be envisaged as a universe in which every point is itself a universe. She described the number of ways a beam of light can travel a closed loop in a two-dimensional universe – an answer you cannot find by staying in your 'home' universe, but only by zooming out and navigating an entire multiverse.

That's further than most of us can aspire to go. But I hope that this book will provide you with your own satisfying journey of mathematical discovery – a way in, at the very least.

Richard Webb, Editor

I
What is mathematics?

What does mathematics consist of? Is it an invention or a discovery? Does it come naturally to us, or must we learn it? When it comes to the true character of mathematics, many questions remain unresolved ...

The pillars of mathematics

For most of us, mathematics means numbers. The manipulation of numbers is certainly where humanity's mathematical journey started. But we have built a formidable, far more extensive edifice on that foundation.

Arithmetic is what we all know: addition, subtraction, division, multiplication and so on. The ability to understand and manipulate numbers in the abstract was the bit of mathematics that we began to develop first, in a formal way as much as six millennia ago. But watertight logical rules of arithmetical manipulation were devised only from the mid-nineteenth century onwards, with the development of set theory.

You can read more about the development of set theory in Chapters 2 and 3 on zero and infinity, and about numbers themselves in Chapters 4 and 5, which deal with the prime numbers, the atoms of the number system, and other particularly intriguing numbers, π, ϕ, e and i.

Probability theory, developed from the seventeenth century onwards, builds on the rules of arithmetic to create its own set of laws for dealing with the chance and uncertainty that is everywhere around us in the world. Originally applied to games of chance, it gained new significance in the twentieth century with the application of statistical methods to analyse large sets of data, and also with the development of quantum theory, which suggests that reality itself is ruled by chance.

Probability and statistics are the subject of Chapter 6, and you will find more on the connection with quantum theory in Chapter 9 on the relationship between numbers and reality.

Beyond the manipulation of numbers is 'higher' mathematics, with three main pillars:

1 **Geometry** is probably the most familiar. It begins with a sense of space: formal geometry codifies the principles for describing how things in space can be related to each other, for example to form a triangle. But it's a static description of things.

2 **Analysis** is the second pillar of higher mathematics. It deals with things that move and change with time. It notably includes integral and differential calculus, together with many other sophisticated variations on the theme.

3 **Algebra** allows us to represent and manipulate knowledge in terms of numbers, symbols and equations, and as such is the broadest pillar of formal higher mathematics. It encompasses esoteric subjects such as group theory (the study of groups, where groups are sets of elements which satisfy certain properties), graph theory (which studies how things are interconnected, such as the computers on the Internet or neurons in the brain) and topology (the mathematics of shapes that can be deformed continuously, without breaking and re-forming them).

Each of these sprawling subjects would be worthy of a book in its own right, but you will gain a flavour of the insights they give and problems they present throughout this book, and particularly in Chapters 7 and 8, dealing with the great unsolved problems of mathematics and the application of mathematics to problems in the everyday world.

Before all that, though, we turn our attention to one of the hardest philosophical questions of mathematics: where does it all come from?

Mathematics: invention or discovery?

Whenever we run to catch a ball or dart through heavy traffic, we do mathematics – entirely unconsciously. That makes sense. The natural world is a complex and unpredictable place. Habitats change, predators strike, food runs out. An organism's survival depends on its ability to make sense of its surroundings, whether by counting down to nightfall, triangulating the quickest way out of danger, or assessing the spots most likely to have food. That means doing mathematics: manipulating numbers, assessing position and movement using trigonometry and calculus, and weighing up probabilities.

This points to a truth that is both profound and difficult to pin down: reality is in some sense mathematical. Karl Friston, a computational neuroscientist and physicist at University College London, observes that there is simplicity, parsimony and symmetry in mathematics. If you were treating it as a language, it would win hands down over all other ways of describing the world.

One immediate consequence is that we are not the only organisms to have 'mathematical' abilities. From dolphins to slime moulds, organisms across the evolutionary tree seem to analyse the world mathematically, deciphering its patterns and regularities in order to survive. If the environment unfolds according to mathematical principles, Friston argues, then the anatomy of the brain must also recapitulate those mathematical principles.

But human brains, with their seemingly unique ability for symbolic representation and abstract thinking, have taken that further. We have made mathematics a conscious activity that, to a greater or lesser extent, must be learned. The exact moment when culture transformed our instinctive senses into a recognizable, conscious mathematical ability is lost in the mists of time, but in the 1970s archaeologists investigating the Border Cave on the western scarp of the Lebombo Mountains in South Africa discovered a series of bones with notches, including the fibula of a baboon etched with 29 such marks. Dated to some 40,000 years ago, they seem to have been an aid to counting – the oldest evidence we have for an emerging conscious understanding for representing and manipulating numbers.

Systems for counting and measuring reached new heights in the fourth millennium BCE, in the sophisticated Mesopotamian culture of the Tigris–Euphrates valleys. Here, in what today we call Iraq, the first consistent symbolic representations of numbers were used to keep track of days, months and years, to measure areas of land and amounts of grain, and perhaps even to record weights. As humans took to the seas and studied the skies, we began developing numerical methods for navigation and for tracking celestial objects.

This conscious mathematics was a product of cultural necessity: an invention that helped to make sense of the world and do things such as trade and travel. With the help of mathematical tools, we have over the past 6,000 years built an immense pyramid of mathematical knowledge. Ancient Greek mathematicians such as Euclid formalized rules of geometry (see Figure 1.1) in around 300 BCE; Hindu and Arabic mathematicians a thousand or so years later began creating the number systems we are familiar with today and developing tools for the symbolic representation and manipulation of numerical quantities: algebra.

But even the great blooming of modern mathematics in the seventeenth-century Age of Enlightenment served only to further our understanding of things within our experience. The calculus of Isaac Newton and Gottfried Leibniz, for example, allowed us to calculate the trajectory of moving bodies on Earth

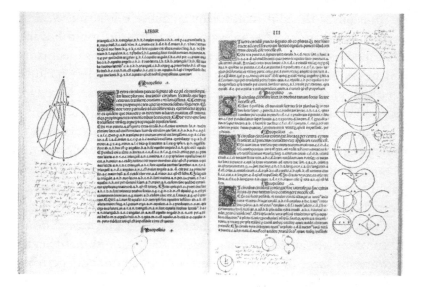

FIGURE 1.1   Euclid's *Elements*, seen here in its first printed edition from 1482, was a seminal primer of geometry.

and in the heavens. The coordinate system invented by René Descartes provided an algebraic representation of geometric shapes. Emerging theories of chance and probability helped us to deal with uncertainty and lack of information.

But mathematics has since expanded into ever more abstract domains, and told us things we could not have hoped to understand by observation alone. As it has done so, it has assumed less and less the character of an invention, a product purely of human brains, and more and more that of a revealed truth, a discovery waiting to be made.

When, for example, at the turn of the twentieth century, the mathematician David Hilbert extended the algebra of conventional 3D space to one with an infinite number of dimensions, it seemed a purely abstract development with little

7

application to the real world. But a couple of decades later it turned out that the state of a quantum particle could best be described using such a 'Hilbert space'. The underlying mathematics remains key to our attempts to make sense of quantum mechanics — a theory of which we have as yet no intuitive physical understanding.

To many physicists today, the success of mathematics as a language to describe reality speaks to a prime role it has in the organization of the universe. Others would not go so far, arguing that we still just invent mathematics to satisfy our need to describe the world differently in different contexts.

Consider the following sequence of events. The most famous of the geometrical axioms that Euclid laid down is that parallel lines never meet. But on the curved surface of the globe, for instance, parallel lines do meet — all lines of longitude meet at the North and South Poles. The exploration by German mathematician Bernhard Riemann and others of such non-Euclidean geometries led to the discovery — or invention — of a rich vein of mathematics that Einstein would use to formulate his general theory of relativity. The warping of space-time by massive bodies in general relativity is dictated by the rules of Riemann's geometry, not Euclid's.

For Andy Clark, a cognitive philosopher at the University of Edinburgh, the universe is filled with all kinds of patterns and regularities and ways of behaving. So any creature that wants to build a mathematics is going to have to build it on top of regularities that are constraining the behaviour of the stuff they encounter. Follow this logic, and if mathematics is an organizing principle, it is one we impose on the world.

Gödel's incompleteness theorems, an ironically rather precise bit of mathematics developed by the Austrian mathematician Kurt Gödel in the 1930s, show that there will always be questions that mathematics will never have the tools to answer (see Chapter 3). That also suggests that it is too early for us to make any sweeping statements about mathematics being a universal truth. We'll return to these thoughts at the end of the book, but in the meantime we are far from what mathematicians would regard as a proof one way or the other.

Our mathematical brains

We all have an innate ability to do a form of mathematics unconsciously, to navigate our way around the world and survive. But the origin of our ability to manipulate numbers is a more intriguing case. Is it learned, or does it tap into something inbuilt? Counting things, after all, has no obvious survival value.

In 1997 the cognitive psychologist Stanislas Dehaene proposed that we are born with a conscious sense of number, in the same way that we are conscious of colours: evolution had endowed humans and other animals with 'numerosity', an ability to immediately perceive the number of objects in some pile of objects. Three red marbles would produce a sense of the number three just as they would produce a sense of the colour red.

Evidence quickly started to accumulate in support of this 'nativist' view of numerical ability, with experiments showing, for example, that six-month-old infants could distinguish between arrays of different numbers of dots. Other studies suggested that humans come with a built-in mental number line – that we instinctively represent numbers spatially, with values increasing from left to right. Experiments appearing to demonstrate that some other animals, from chimps to chickens, can distinguish small numbers seemed to provide supplementary evidence.

The development of numbers

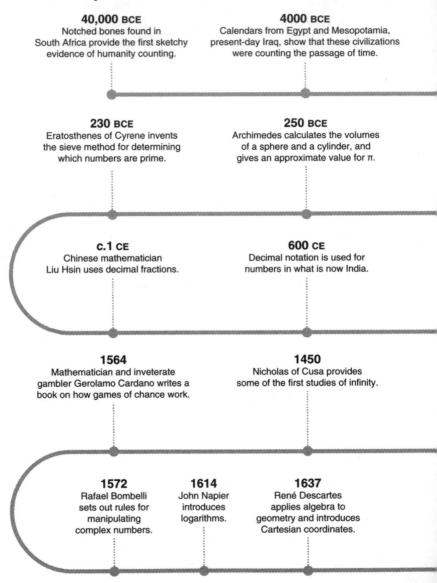

40,000 BCE
Notched bones found in
South Africa provide the first sketchy
evidence of humanity counting.

4000 BCE
Calendars from Egypt and Mesopotamia,
present-day Iraq, show that these civilizations
were counting the passage of time.

230 BCE
Eratosthenes of Cyrene invents
the sieve method for determining
which numbers are prime.

250 BCE
Archimedes calculates the volumes
of a sphere and a cylinder, and
gives an approximate value for π.

c.1 CE
Chinese mathematician
Liu Hsin uses decimal fractions.

600 CE
Decimal notation is used for
numbers in what is now India.

1564
Mathematician and inveterate
gambler Gerolamo Cardano writes a
book on how games of chance work.

1450
Nicholas of Cusa provides
some of the first studies of infinity.

1572
Rafael Bombelli
sets out rules for
manipulating
complex numbers.

1614
John Napier
introduces
logarithms.

1637
René Descartes
applies algebra to
geometry and introduces
Cartesian coordinates.

3000 BCE
The first recognizable number system emerges in Mesopotamia.

1750 BCE
The Babylonians are solving linear and quadratic equations, and compiling tables of square and cube roots.

300 BCE
Euclid writes *The Elements,* among other things a comprehensive primer of geometry.

530 BCE
Pythagoras of Samos proves his eponymous theorem about the length of sides of a right-angled triangle.

628 CE
The Hindu mathematician Brahmagupta writes *Brahmasphutasiddhanta,* introducing negative numbers and the first numerical zero.

810 CE
The Arabic mathematician al-Khwārizmī introduces the term 'algebra' and gives his own name to 'algorithm'.

1202
Fibonacci (Leonardo of Pisa) writes *Liber abaci,* introducing Arabic arithmetic and algebra to Europeans.

1072
Al-Khayyami (Omar Khayyam) calculates the length of the year to be 365.24219858156 days – a stunningly accurate value.

1647
Pierre de Fermat writes down his mystery last theorem – which remains unsolved until 1994.

1664
Fermat and Blaise Pascal exchange letters in which they begin to sketch out the laws of probability.

The development of numbers

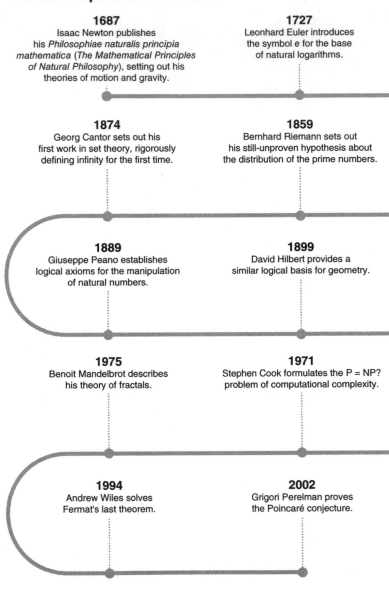

1687
Isaac Newton publishes
his *Philosophiae naturalis principia
mathematica* (*The Mathematical Principles
of Natural Philosophy*), setting out his
theories of motion and gravity.

1727
Leonhard Euler introduces
the symbol *e* for the base
of natural logarithms.

1874
Georg Cantor sets out his
first work in set theory, rigorously
defining infinity for the first time.

1859
Bernhard Riemann sets out
his still-unproven hypothesis about
the distribution of the prime numbers.

1889
Giuseppe Peano establishes
logical axioms for the manipulation
of natural numbers.

1899
David Hilbert provides a
similar logical basis for geometry.

1975
Benoit Mandelbrot describes
his theory of fractals.

1971
Stephen Cook formulates the P = NP?
problem of computational complexity.

1994
Andrew Wiles solves
Fermat's last theorem.

2002
Grigori Perelman proves
the Poincaré conjecture.

1742
Christian Goldbach sets
out his conjecture about the
construction of prime numbers.
It remains unproved.

1763
Thomas Bayes's theory
of probability is published
posthumously.

1844
Joseph Liouville finds the
first transcendental numbers.

1843
William Rowan Hamilton discovers
quaternions, four-dimensional
complex numbers.

1901
Bertrand Russell uncovers a
logical paradox undermining
set theory.

1904
Henri Poincaré formulates his
conjecture concerning the geometry
of higher-dimensional shapes.

1947
George Dantzig introduces his
simplex algorithm, vastly simplifying
optimization problems.

1931
Kurt Gödel shows in his
incompleteness theorems that
any system of mathematical logic
is bound to be incomplete.

But, before long, some researchers grew uncomfortable with the conclusions of these studies. The subjects might, for example, be distinguishing arrays of dots based not on number but on other attributes such as their spatial distribution or area of coverage. Tali Leibovich of the University of Haifa in Israel points out that it makes sense that we would have evolved to assess these things: if you are hunting or being hunted, you need to act quickly, which would mean using all available cues.

Soon, a different hypothesis emerged: instead of being born with an innate sense of number, we are born with a sense for quantities such as size and density that are correlated with the numbers of things, and our conscious mathematical ability builds on this. 'It takes time and experience to develop and understand this correlation,' says Leibovich.

More refined tests in children tend to support this view. Children younger than about four years of age cannot understand that five oranges and five watermelons have something in common: the number five. To them, a bunch of watermelons simply represents more 'stuff' than the same number of oranges.

Observations of different human cultures provide additional evidence. The Yupno people of Papua New Guinea have a complex language including subtle demonstrative pronouns indicating whether something is physically higher or lower than the speaker, how many things there are, and exactly how near or far they are (English, by contrast, has only four demonstratives: this, that, these and those). But the Yupno do not use the supposedly universal mental number line, nor do they have comparatives in their language to say that something is bigger or smaller. Rafael Núñez of the University of California San Diego, who has studied the Yupno culture for many years, says that it is not just the lack of exact quantification: they also lack grammatical properties to support something as simple as a comparison of size or weight.

Núñez points also to a study of 189 Aboriginal Australian languages, of which three-quarters were found to have no words for numbers above three or four, while a further 21 went no further than five. To Núñez, this suggests that exact numerosity is not innate, but a cultural trait that emerges when circumstances, such as agriculture and trading, demand it. 'Hundreds of thousands of humans who have language, sometimes very complicated and sophisticated language, don't have exact quantification,' he says.

And some people definitely learn numerosity better than others. In 2016 Dehaene reported the results of scanning the brains of 15 professional mathematicians and 15 non-mathematicians of equal academic ability. They found a network of brain regions involved in mathematical thought activated when mathematicians reflected on problems in algebra, geometry and topology, but not when they were thinking about unrelated areas. No such distinction was visible in the non-mathematicians.

Crucially, this 'maths network' does not overlap with brain regions involved in language, suggesting that once the mathematically able develop and learn the language for manipulating symbols, they start thinking in ways that do not involve normal language. For Friston, it is as though these people are able to download an intuition into another world, the world of mathematics, and stand back and let it talk back to them. That ability probably leans on many other things: language to communicate concepts, working memory to hold and manipulate concepts, and even cognitive control to overcome innate biases in our brains.

When mathematics goes wrong

As a product of the random processes of natural selection, our unconscious mathematical models of the world are, seen from the point of view of conscious mathematics, not perfect.

Sometimes they prioritize keeping us alive at the expense of accuracy – a source of all sorts of common mathematical pitfalls.

This is one reason, for instance, why we find it so difficult to assess probabilities. We tend to inflate our estimate of risks – better safe than sorry – and see patterns where there are none. That lies behind things such as the gambler's fallacy, the mistaken belief that, if the roulette wheel keeps landing on red, a bet on black is the safer one to make (see Chapter 8).

Or take the Weber–Fechner effect, which governs our response to external stimuli. It states that our ability to discriminate between sensations diminishes as their magnitude increases. While a 2-kg weight can easily be distinguished from a 1-kg weight, for example, weights of 22 kg and 21 kg are harder to tell apart. Similar things apply to the brightness of lights, the volume of a sound and even the number of objects you can see.

Experiments show that other animals share these inbuilt flaws – but as yet only we humans have developed the ability to identify and potentially overcome these flaws, in the form of conscious mathematics.

How to think about mathematics

How do its practitioners get themselves thinking mathematically? Ian Stewart of the University of Warwick, UK, sees his subject as akin to a language – but one that, thanks to its inbuilt logic, writes itself. 'You can start writing things down without knowing exactly what they are, and the language makes suggestions to you,' he says. Master enough of the basics, and you rapidly enter what sports players call 'the zone'. In this state, Stewart has found, things get much easier and you are propelled along.

But what if you lack such a maths drive? It's wrong to think that it's all down to talent, says mathematician and writer Alex Bellos: even the best exponents can take decades to master their craft. He believes one of the reasons people don't understand maths is that they simply don't have enough time.

Sketching a picture of the problem helps. Take negative numbers. Five sheep are easy enough to envisage, but minus five sheep are really difficult to get your head around. It was only when someone had the bright idea of arranging all the existing numbers 0,1,2,3 … on a line that it became obvious where the negative numbers fitted in. Similarly, complex numbers only really took off with the advent of a 'complex plane' in which to depict them (see Chapter 5).

Analogies also help. Stewart's advice is that, if thinking about ellipses oppresses you, consider instead a circle that has been squashed and work from there. Overall, contrary to the impression of mathematics as a discipline of iron logic, the best way to attack a problem of any sort is often to get a brief overview of it, skip over anything you cannot work out and then go back and fill in the details. 'A lot of mathematicians say it's important to be able to think vaguely,' says Stewart.

Interview: Inspiration from Rubik's cube

Mathematical genius is measured in Fields Medals. Two, three or four medals are awarded once every four years to mathematicians under the age of 40. Together with the Abel Prize, they are regarded as the highest honour in mathematics. Manjul Bhargava – one of the youngest people to be made a full professor at Princeton University, aged 28, in 2003 – was awarded one in 2014. Here he reveals the unusual ways his mathematical brain works.

Did the Fields Medal mean more to you than any other award you have won?

Any award is a milestone, which encourages one to go further. I don't know that I think of any award as meaning more to me personally than any other. The mathematics that led to the medal was far more exciting to me than the medal itself.

The award citation says that you were inspired to extend Gauss's law of composition in an unusual way. What does that mean, and what did you do?

Gauss's law says that you can compose two quadratic forms, which you can think of as a square of numbers, to get a third square. I was in California in the summer of 1998, and I had a $2 \times 2 \times 2$ mini Rubik's cube. I was just visualizing putting numbers on each of the corners, and I saw these binary quadratic forms coming out, three of them. I just sat down and wrote out the relations between them. It was a great day!

Have any of your other discoveries had unusual origins?

I do tend to think about things very visually, and the Rubik's cube is a concrete example of that visual approach. But that one is probably the most unusual and unexpected origin of all.

You have proved several theorems. Do you have a favourite?

Mathematicians often say that choosing a favourite theorem is like choosing one's favourite child. Although I don't yet have any children, I understand the sentiment. I enjoyed working on all the theorems I have proved.

Are there any mathematicians, living or dead, whom you have particularly looked up to?

My mother [Mira Bhargava, a mathematician at Hofstra University in Hempstead, New York] has been a source of inspiration to me from the very beginning. She was always there to answer my questions, to encourage and support me, and she taught me how much the human mind is capable of.

2
Zero

It is the embodiment of nothing, but is it something? We start our exploration of numbers with a concept mathematicians have wrestled with for centuries. Is zero a number, or isn't it?

An untameable nothing

Zero does not behave like a normal number. Add any two numbers together and you will get a different number – unless one of them is zero. Multiply any number by zero, and the result is only ever zero – again, something that happens with no other number. And don't even try to divide a number by zero: mathematicians generally call that result 'undefined' because, if you do it, pandemonium ensues (see 'Proving 1 = 2' below).

These quirks are partly why zero existed as a symbol long before it was accepted as a number. This symbolic zero is familiar as a 'placeholder' in our positional numerical notation based on powers of 10. Take the string of digits '2018', for example. It has a value equal to $2 \times 10^3 + 0 \times 10^2 + 1 \times 10^1 + 8$. Zero's role is pivotal: were it not there, we might easily mistake 2018 for 218, or perhaps 20018, and our calculations could be out by hundreds or thousands.

The first positional number system was used to calculate the passage of the seasons and the years in Babylonia, part of Mesopotamia, modern-day Iraq, from around 1800 BCE onwards. Its base was not 10, but 60, and it had just two symbols, for 1 and 10, that were roughly clumped together to give the value of a number. Counting with such a system must have been a headache, not least because the absence of any power of 60 in the transcription of any number was marked not by a symbol but (if you were lucky) just by a gap. Only in around 300 BCE did a third symbol, a curious confection of two left-slanting arrows, start to fill missing places in the stargazers' calculations.

This was the world's first symbol for zero. Some seven centuries later, zero the symbol was invented on the other side of the world for a second time, when Mayan priest-astronomers in Central America began to use a snail-shell-like symbol to fill gaps in the positional number system they used to calculate their calendar. But neither the Babylonians nor the Mayans took the next step: to treat zero as a number in its own right that could be manipulated or even be the result of a calculation.

Proving that 1 = 2

Some simple algebra can be used to prove that 1 = 2, with the help of an illegal mathematical operation.

1. Let $a = b$
2. Multiply both sides by a: $\quad\quad\quad a^2 = ab$
3. Add a^2 to both sides: $\quad\quad\quad 2a^2 = a^2 + ab$
4. Subtract 2ab from both sides: $\quad 2a^2 - 2ab = a^2 - ab$
5. This can be written as $\quad\quad\quad 2(a^2 - ab) = 1\,(a^2 - ab)$
6. Dividing both sides by
   $(a^2 - ab)$ gives $\quad\quad\quad\quad\quad 2 = 1.$

Where did the proof go wrong? In step 1, we defined $a = b$, so $a^2 - ab = 0$. This invalidates step 6, making it equivalent to dividing by 0 – a mathematical no-no.

Though otherwise far-seeing mathematicians, the ancient Greeks did not fare much better. Greek thought was wedded to the idea that numbers expressed geometrical shapes; and what shape would correspond to something that was not there? It could only be the total absence of something, the void – a concept that the dominant cosmology of the time had banished. For the Greeks, and successor Christian civilizations in Europe, zero was a godless concept. So although they often used the Babylonian placeholder zero, for example in astronomical calculations, they, too, avoided zero the number.

Zero the number

Eastern philosophies had no such qualms, so the next great staging post in zero's journey was not to Babylon's west, but to its east. It is found in *Brahmasphutasiddhanta*, a treatise on the relationship of mathematics to the physical world written in India around 628 CE by the astronomer Brahmagupta. He was the first person we know of who treated numbers as purely abstract quantities separate from any physical or geometrical reality. That allowed him to consider unorthodox questions such as what happens when you subtract from one number a number of greater size. In geometrical terms this is a nonsense: what area is left when a larger area is subtracted?

Brahmagupta instead envisaged numbers on a continuous line stretching as far as you could see in both directions, showing both positive and negative numbers, and gave rules for how to manipulate these quantities that we would recognize today. Sitting in the middle of this line, a distinct point along it at the threshold between the positive and negative worlds, was what he called *sūnya*, the nothingness.

It was not long before this new number was unified with zero the symbol (see Figure 2.1). Recent dating of a manuscript held in the Bodleian Library in Oxford, UK, suggests that as early as the third or fourth century CE Hindu mathematicians had been using a squashed egg symbol, recognizably close to our own zero, as a placeholder. Brahmagupta's innovation soon made this placeholder a full member of a modern, 'dynamic' positional number system with a full count to digits up to its base, from 0 to 9.

Here, zero assumes almost unannounced a new guise: it becomes a mathematical tool that brings to bear the power of the system's base. Add a zero to the end of a decimal number string, and we multiply it by 10: 2018 becomes 20180. Sum

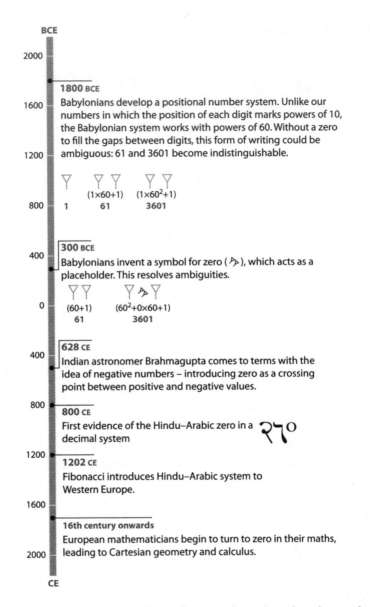

FIGURE 2.1 Zero is crucial for mathematics, but it has taken thousands of years for its importance to be recognized.

two or more numbers so that the total of a column ticks over from 9 to 10, and we 'carry the one' and leave a zero to ensure the right answer.

The simplicity of such algorithms is the source of our system's supple muscularity in manipulating numbers. Soon it spawned a new way of doing mathematics among Hindu and Arabic mathematicians: algebra.

News of these innovations took a long time to filter through to Europe. In 1202 a young Italian, Leonardo of Pisa – better remembered as Fibonacci – published a book, *Liber Abaci*, in which he presented details of the Arabic counting system he had encountered on a journey to North Africa. He demonstrated the superiority of this notation over the abacus, and the then prevalent, non-positional, system of Roman numerals, for the deft performance of complex calculations.

While merchants and bankers were quickly convinced of the Hindu–Arabic system's usefulness (see Figure 2.2), the governing authorities were less enamoured. In 1299 the city of Florence banned the use of the Hindu–Arabic numerals, including zero, considering the ability to inflate a number's value hugely simply by adding a digit on the end to be an open invitation to fraud.

Eventually, though, zero won through. The seventeenth century saw the invention by Frenchman René Descartes of the Cartesian coordinate system, which married algebra and geometry to give every geometrical shape a new symbolic representation with zero at its centre. Soon afterwards, Isaac Newton and Gottfried Leibniz invented the new tool of calculus, which showed that you had first to appreciate how zero merged into the infinitesimally small to explain how anything in the cosmos could change its position at all – a star, a planet, a hare overtaking a tortoise.

FIGURE 2.2 In this woodcut from the *Margarita philosophica* (*Pearl of Wisdom*), a fifteenth-century encyclopaedia compiled by the German Gregor Reitsch, the Muse of Arithmetic is depicted smiling on Boethius (left) calculating with Hindu–Arabic numerals, while Pythagoras (right) struggles with the old way of using an abacus.

A better understanding of zero became the fuse of the Scientific Revolution that followed. But it was only in the nineteenth century that we came to realize just how essential zero is to mathematics itself. The key was the advent of set theory.

Explaining numbers: set theory

In the late 1800s, while most mathematicians were busy adding a nice piece of furniture, a new room, even an entire storey to the growing mathematical edifice, a group of worrywarts started to fret about the cellar. Innovations of the time such as non-Euclidean geometry were all very well – but were the underpinnings sound? To prove they were, a basic idea needed sorting out that no one really understood. Numbers.

By this stage, using numbers was not the problem. The big question was what they were. You can show someone two sheep, two coins, two albatrosses, two galaxies. But can you show them two? It is important to see that the symbol '2' is a notation, not the number itself: many cultures use a different symbol. The same is true for the word 'two': in other languages it might be *deux* or *zwei* or *futatsu*. For thousands of years humans had been using numbers to great effect, but suddenly a few deep thinkers realized that no one had a clue what they were.

An answer emerged from two different lines of thought: mathematical logic, and Fourier analysis, in which a complex waveform is represented as a combination of simple sine waves. These two areas converged on one idea: sets. A set is a collection of mathematical objects – numbers, shapes, functions, networks, whatever. It is defined by listing or characterizing its members. 'The set with members 2, 4, 6, 8' and 'the set of even integers between 1 and 9' both define the same set, which can be written as $\{2, 4, 6, 8\}$.

Around 1880 Georg Cantor developed an extensive theory of sets. He had been trying to sort out some technical issues in Fourier analysis related to discontinuities – places where a waveform makes sudden jumps. His answer involved the structure of the set of discontinuities. It was not the individual discontinuities that mattered; it was the whole class of discontinuities.

One thing led to another. Cantor devised a way to count how many members a set has, by matching it in a one-to-one fashion with a standard set. Suppose, for example, the set is {Doc, Grumpy, Happy, Sleepy, Bashful, Sneezy, Dopey}. To count them, we chant '1, 2, 3 …' while working along the list: Doc (1), Grumpy (2), Happy (3), Sleepy (4), Bashful (5), Sneezy (6), Dopey (7). Right: seven dwarfs. We can do the same with the days of the week: Monday (1), Tuesday (2), Wednesday (3), Thursday (4), Friday (5), Saturday (6), Sunday (7).

Another mathematician of the time, Gottlob Frege, picked up on Cantor's ideas and thought they could solve the big philosophical problem of numbers. The way to define them, he believed, was through the deceptively simple process of counting.

What do we count? A collection of things – a set. How do we count it? By matching the things in the set with a standard set of known size. The next step was simple but devastating: throw away the numbers. You could use the dwarfs to count the days of the week. Just set up the correspondence: Monday (Doc), Tuesday (Grumpy) … Sunday (Dopey). There are Dopey days in the week. It's a perfectly reasonable alternative number system. It doesn't (yet) tell us what a number is, but it gives a way to define 'same number'. The number of days equals the number of dwarfs, not because both are seven, but because you can match days to dwarfs.

What, then, is a number? Mathematical logicians realized that to define the number 2, you need to construct a standard set which intuitively has two members. To define 3, use a standard set with three numbers, and so on. That just moved the question on, however: what standard set should you use? It should be unique, and its structure should correspond to the process of counting. Enter zero – and the empty set.

Georg Cantor's heaven and hell

Born in 1845, Georg Cantor was the father of set theory, the underpinning of all modern considerations of number. But his groundbreaking work, especially his investigations of the nature of infinity, were not always well received. His contemporary Henri Poincaré is said to have described set theory as a 'disease from which one has recovered'. Some Christian theologians saw his work on infinity as a direct challenge to conceptions of God.

Cantor, himself devoutly religious, seemed to suffer from the paradoxes and challenges he encountered in his work, and experienced mental breakdowns with increasing frequency. In the final decades of his life he abandoned mathematics entirely, concentrating instead on trying to prove that Shakespeare's plays were written by the philosopher Francis Bacon and that Jesus was the son of Joseph of Arimathea. By that stage, however, the set theory he had created was well established. In 1926 the mathematician David Hilbert said that 'no one will expel us from the paradise that Cantor has created'.

The empty set

If zero is a number, it ought to count the members of a set. But which set? Well, it has to be a set with no members. These aren't hard to think of: 'the set of all mice weighing 20 tonnes', perhaps. There is also a mathematical set with no members: the empty set. It is unique, because all empty sets have exactly the same members: none. Its symbol, introduced in 1939 by a group of mathematicians that went by the pseudonym Nicolas Bourbaki, is ∅. Set theory needs ∅ for the same reason that arithmetic needs 0: things are a lot simpler if you include it. In fact, we can define the number 0 as the empty set.

What about the number 1? Intuitively, we need a set with exactly one member. Something unique. Well, the empty set is unique. So we define 1 to be the set whose only member is the empty set: in symbols {∅}. This is not the same as the empty set because it has one member, whereas the empty set has none. Agreed, that member happens to be the empty set, but there is one of it. Think of a set as a paper bag containing its members. The empty set is an empty paper bag. The set whose only member is the empty set is a paper bag containing an empty paper bag. Which is different: it has a bag in it (see Figure 2.3).

The key step is to define the number 2. We need a uniquely defined set with two members. So why not use the only two sets we have mentioned so far: ∅ and {∅}? We therefore define 2 to be the set {∅, {∅}}. Which, thanks to our definitions, is the same as {0, 1}.

Now a pattern emerges. Define 3 as {0, 1, 2}, a set with three members, all of them already defined. Then 4 is {0, 1, 2, 3}, 5 is {0, 1, 2, 3, 4} and so on. Everything traces back to the empty set: for instance, 3 is {∅, {∅}, {∅, {∅}}} and 4 is {∅, {∅}, {∅, {∅}}, {∅, {∅}, {∅, {∅}}}}. You don't want to see what the number of dwarfs looks like.

The building materials here are abstractions: the empty set and the act of forming a set by listing its members. But the way these sets relate to each other leads to a well-defined

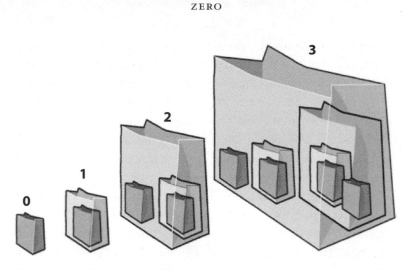

FIGURE 2.3 The empty set has no member, like an empty paper bag. But by putting the empty paper bag in a larger paper bag you can form big and bigger sets – the basis of our definition of number.

construction for the number system, in which each number is a specific set that intuitively has that number of members. The story doesn't stop there. Once you have defined the positive whole numbers, similar set-theoretic trickery defines negative numbers, fractions, real numbers (infinite decimals), complex numbers and so on. It's the dreadful secret of mathematics: it's all based on nothing.

3
Infinity

At the other end of the number scale from zero is infinity. With it, we made a monster. Our minds demand that it should exist, only to rapidly melt at the consequences of a concept that is too big for our brains.

What is infinity?

Infinity has a pre-programmed boggle factor. Mathematically, it started off as a way of expressing the fact that some things, like counting, have no obvious end. Count to 146 and there's 147; count to a trillion and say hello to a trillion and one. There are two ways of dealing with this. You can be cautious and say that there is no largest number – there is just a 'potential' for infinity. Or you can be bolder and say that there are infinitely many numbers, treating infinity as an actual quantity with properties all its own.

Only in the late nineteenth century did mathematicians plump for actual infinity. As with elucidating the true value of zero, the key was set theory (see Chapter 2). The set of whole numbers, 1,2,3,4 ..., for example, is a well-defined and unique object, and it has a size: infinity.

But now write out the squares of those numbers: 1,4,9,16 ... This sequence gets bigger a lot faster, so it must reach infinity faster, right? Not so. As Galileo recognized back in the seventeenth century, every whole number has a square, so there are as many square numbers as whole numbers – infinitely many.

There are in fact very many of these 'countable' infinities related to the countable natural numbers. You can show that they all have the same size – the basis of the seemingly paradoxical problem of Hilbert's infinite hotel (see 'Hilbert's hotel' below). It is within countable infinity that the action of arithmetic takes place.

So infinity is infinity is infinity – except that it isn't. Now look at the real numbers: the whole numbers plus all the rational and irrational numbers in between (1.5, π, the square root of 2 and so on). Asking how many of these there are is equivalent to asking the question 'How many points are there on a line?' A line is perfectly straight and smooth, with no holes or gaps; it contains infinitely many points, so the answer is again 'infinitely many'.

As the great set theorist Georg Cantor showed, however, this 'continuum' infinity is a bigger number than countable infinity. In fact, it is merely a step in a staircase leading to ever-higher levels of infinities stretching up as far as, well, infinity. The relationship between the countable, the continuum and all the other types of infinity is one of the greatest unanswered questions of mathematics.

Hilbert's hotel

David Hilbert's paradox of the Grand Hotel tells us that infinite sets do not work as we might intuitively think.

Imagine a hotel with a (countably) infinite number of rooms, each of which is occupied. A coachload of 50 more guests arrives. Where can they be accommodated? In an infinite hotel, it's not a problem: you move each guest already in the hotel up 50 rooms, and accommodate the new guests in rooms 1–50.

In fact, using a slightly different filling algorithm, you can accommodate an infinitely large coachload of new guests: just move the resident of room 1 to room 2, the resident of room 2 to room 4, and so on, moving people from room n to room $2n$. Then put the newcomers in the odd-numbered rooms, of which there are infinitely many, all now free.

Hilbert's seeming paradox tells us that, tempting as it might be to think that there are half as many even numbers as there are whole numbers in total, the two collections can in fact be paired off exactly. Indeed, every set of

---

whole numbers is either finite or countably infinite, and the countably infinite sets are all the same size. It doesn't stop there. With a little more mathematical effort you can show that Hilbert's hotel can accommodate an infinite number of infinite coachloads of new guests, and so on … ad infinitum.

The continuum hypothesis

When David Hilbert left the podium at the Sorbonne in Paris on 8 August 1900, few of the assembled delegates seemed overly impressed. According to one contemporary report, the discussion following the great mathematician's address to the second International Congress of Mathematicians was 'rather desultory'. Passions were more inflamed by a subsequent debate on whether Esperanto should be adopted as the working language of mathematics.

Yet Hilbert's address set the mathematical agenda for the twentieth century, crystallizing into a list of 23 crucial unanswered questions. Today many of these problems have been resolved, including how to pack spheres to make best use of the available space (see Chapter 7). Others, such as the Riemann hypothesis, which concerns how the prime numbers are distributed, have seen little or no progress. But the first item on Hilbert's list stands out for the sheer oddness of the answer supplied by generations of mathematicians since: that mathematics is not equipped to provide an answer.

First formulated by Georg Cantor, it is known as the continuum hypothesis. It states that there is no intermediate level of infinity between the countable level, encompassing the whole numbers, and the continuum level, encompassing the real numbers.

Attempts to prove or disprove the continuum hypothesis involve analysing all possible infinite subsets of the real numbers. If every such set is either countable or has the same size as the full continuum, then the continuum hypothesis is correct. Conversely, even one subset of intermediate size would render it false.

Cantor could not prove his hypothesis because, as the British mathematician and philosopher Bertrand Russell showed, he had been too hasty. Although his conclusions about infinity were sound, the logical basis of his set theory was flawed, resting on an informal and ultimately paradoxical conception of what sets are.

Russell's paradox

Formulated by Bertrand Russell in 1901, this showed up a flaw in Georg Cantor's original formulation of set theory that only later refinements were able to correct.

Imagine a set, R, defined as the set of all sets that do not contain themselves. If R is not a member of itself, then its definition dictates it is a member of itself. But it is not a member of itself. So which is it?

The basic problem of self-reference raised by Russell's paradox – logical statements or objects that refer to themselves – bedevils all systems of logic. It is also at the heart of the far earlier liar paradox embodied by the sentence 'This sentence is false'. Is that true or false? Kurt Gödel would later show in his incompleteness theorems that self-reference represents a fundamental problem for mathematics.

It was not until 1922 that two German mathematicians, Ernst Zermelo and Abraham Fraenkel, devised a series of rules for manipulating sets that was seemingly robust enough to support Cantor's tower of infinities and stabilize the foundations of mathematics. Unfortunately, though, these rules delivered no clear answer to the continuum hypothesis. In fact, that and other developments at the time seemed strongly to suggest there might even not be an answer.

The axiom of choice

The immediate stumbling block was a rule known as the 'axiom of choice'. It was not part of Zermelo and Fraenkel's original rules, but it was soon bolted on when it became clear that some essential mathematics, such as the ability to compare different sizes of infinity, would be impossible without it.

The axiom of choice states that, if you have a collection of sets, you can always form a new set by choosing one object from each of them. That sounds anodyne, but it comes with a sting. The Polish mathematicians Stefan Banach and Alfred Tarski soon showed how the axiom could be used to divide the set of points defining a spherical ball into six subsets, which could then be slid around to produce two balls of the same size as the original. That was a symptom of a fundamental problem: the axiom of choice allowed an assortment of peculiarly perverse sets of real numbers to exist.

This news came at a time when the concept of 'unprovability' was just coming into vogue. In 1931 the Austrian logician Kurt Gödel proved his notorious incompleteness theorems, showing that even with the most tightly knit basic rules there will always be statements about sets or numbers, such as the axiom of choice or the continuum hypothesis, that mathematics can neither prove nor disprove.

FIGURE 3.1 Many great figures in mathematics contributed to the development of set theory, and with it the logical underpinnings of arithmetic, in the late nineteenth and early twentieth centuries.

Georg Cantor (1845–1918) Gottlob Frege (1848–1925)

1874 Invents the idea of a 'set' as a mathematical collection of objects, and establishes that infinite sets come in different sizes.

1893 Establishes that, if numbers are interpreted as measuring the sizes of sets, Peano's rules are obeyed.

Alfted North Whitehead (1861–1947)

1910–13 Whitehead and Russell succeed in revising Frege's ideas to avoid Russell's paradox in their monumental work *Principia Mathematica*.

Ernst Zermelo (1871–1953) Abraham Fraenkel (1891–1965)

1922 The German mathematicians put set theory on a firm logical footing, with Cantor's infinite sets at its heart.

Alan Turing (1912–54)

1937 Using his theoretical Turing machine, later the basis of the digital computer, Turing shows that Hilbert's idea of universal arithmetical computability will never be fulfilled.

Giuseppe Peano (1858–1932)

1889 The Italian writes down the standard logical rules that underlie arithmetic, kick-starting the study of maths' underlying logic.

Bertrand Russell (1872–1970)

1901 The British mathematician and philosopher produces a notorious paradox showing that the naive idea of sets leads to contradictions within Cantor and Frege's systems.

David Hilbert (1862–1943)

1920 Hilbert proposes a programme to establish that the rules of arithmetic are complete, consistent and computable.

Kurt Gödel (1906–78)

1931 The Austrian logician establishes in his incompleteness theorems that Hilbert's hopes for completeness and consistency in arithmetic are mutually incompatible.

1870 1880 1890 1900 1910 1920 1930 1940

Gödel incompleteness

Kurt Gödel was just 25 years old when he published his two incompleteness theorems in 1931. They formalized the kind of logical inconsistencies in set theory uncovered by Bertrand Russell and others into a general statement about the limits of mathematics.

The first incompleteness theorem states that any self-consistent system of logical axioms that can describe the arithmetic of the natural numbers will leave statements about the natural numbers that may be true but cannot be proved. The second incompleteness statement goes on to say that such a system of axioms cannot be used to prove its own consistency. To prove consistency requires building a bigger logical structure around it. But that, too, will suffer from its own incompleteness.

At the same time, though, Gödel had a crazy-sounding hunch about how you might fill in most of these cracks in mathematics' underlying logical structure: you simply build more levels of infinity on top of it. He proved his point in 1938. By starting from a simple conception of sets compatible with Zermelo and Fraenkel's rules and carefully tailoring its infinite superstructure, he created a mathematical environment in which both the axiom of choice and the continuum hypothesis are simultaneously true. He dubbed his new world the 'constructible universe' – or simply 'L'.

Since then, mathematicians have discovered a vast diversity of models of set theory. Some are 'L-type worlds' with superstructures like Gödel's L, differing only in the range of extra levels of infinity they contain; others have wildly varying architectural styles with completely different levels and infinite staircases leading in all sorts of directions.

For most purposes, life within these structures is the same: everyday mathematics does not generally differ between them, and nor do the laws of physics. But the existence of this mathematical 'multiverse' also seemed to dash any notion of ever getting to grips with the continuum hypothesis. As mathematician Paul Cohen was able to show in the 1960s, in some logically possible worlds, the hypothesis is true and there is no intermediate level of infinity between the countable and the continuum; in others, there is one; in still others, there are infinitely many. With mathematical logic as we know it, there is no way of finding out which world we are in and whether the continuum hypothesis is true or not.

The infinity man

Set theorist Hugh Woodin of Harvard University knows better than most how to think about a concept as mindblowing as infinity. He has a whole level of infinity named after him, a particularly vertiginous level populated with numbers known as Woodin cardinals. 'They are so large you can't deduce their existence,' he says.

Such infinities are the ultimate abstraction: although you can manipulate them logically, you can't write formulae incorporating them or devise computer programs to test predictions about them. Woodin's infinities serve to set aside some of the inconsistencies in set theory uncovered by Kurt Gödel and others. 'Set theory is riddled with unsolvability. Almost any question you want to ask is unsolvable,' he says. He has been working on a new logical superstructure for mathematics that, in tribute to Kurt Gödel's logical world called L, he has dubbed 'ultimate L'.

Ultimate L implies that Cantor's continuum hypothesis is true, so there is no intermediate level between countable and continuum infinity. But it does not rest there. Its wide, airy space allows extra steps to be bolted to the top of the infinite staircase as necessary to fill in gaps below, making good on Gödel's hunch about using infinities to root out the unsolvability that riddles mathematics. Gödel's incompleteness theorem would not be dead, but you could chase it as far as you pleased up the staircase into the infinite attic of mathematics.

Is infinity real?

Infinity is essential to the logical structure of mathematics. In practical terms, too, there's very little that works smoothly without manipulating the infinite and its obverse, the infinitesimal. In geometry, for example, defining a perfect circle requires the infinite digits of π, while mathematical functions such as sines and cosines that relate angles to the ratios of two line lengths are defined by infinite numbers of terms. In mechanics, calculating smooth motions requires chopping time into infinitesimally small chunks.

But is any of this truly real? Take the infinite set of whole numbers: you could never actually write all of them out – you would die before you got to the end. Even if someone else took over, a finite universe would eventually run out of paper on which to write them down – and of information to encode them with (see Chapter 9).

That is reason enough to tread with caution, argues mathematician Norman Wildberger of the University of New South Wales in Sydney, Australia. He points out that, for most of history, infinity was kept at arm's length. For greats of the subject from Aristotle to Newton, the only infinity was potential infinity, the sort that allows us to add 1 to any number without fear of hitting the end of the number line, but is never actually reached. That is a long way from accepting an infinity that has already been reached and conveniently packaged as a mathematical entity.

For Wildberger, the problems of set theory should be tackled not with more infinity, but with less. This is because many of the serious logical weaknesses in modern mathematics are associated in one way or another with infinite sets of real numbers. Since the 1990s Wildberger has been working on a new, infinity-free version of trigonometry and Euclidean geometry. His 'rational geometry' aims to avoid these infinities, replacing angles, for example, with a 'spread', a finite quantity extracted from mathematical vectors representing two lines in space.

Doron Zeilberger of Rutgers University in Piscataway, New Jersey, wants to take an even more radical approach and dispose of potential infinity as well. Forget everything you thought you knew about mathematics: there is a largest number. Start at 1 and just keep on counting and eventually you will hit a number you cannot exceed – a kind of speed of light for mathematics.

This raises a host of questions. How big is the biggest number? Zeilberger says that it is so big we could never reach it, so instead he's given it a symbol, N_0. What happens if you add 1 to it? Zeilberger's answer comes by analogy to a computer processor. Every computer has a largest integer number that it can handle: exceed it, and you will either get an 'overflow error' or the processor will reset the number to zero.

So far, finitist mathematics has received most attention from computer scientists and robotics researchers, who work with finite forms of mathematics as a matter of course. Finite computer processors cannot approximate the potentially infinite digits of real numbers. Instead, they use floating-point arithmetic, a form of scientific notation that allows the computer to drop digits and save on memory without losing a number's overall scope. As early as 1969 Konrad Zuse, a German engineer and one of the pioneers of floating-point arithmetic, argued that the universe itself is a digital computer – one with no room for infinity.

4
Prime numbers

Prime numbers are the building blocks from which all other numbers are made. They hold many mysteries — understand them, and we can understand some of the deepest problems of mathematics.

Why are prime numbers important?

The prime numbers are the atoms of the number system, those numbers bigger than 1 divisible only by 1 and themselves. The sequence starts 2, 3, 5, 7, 11, 13, 17, 19 … and is infinitely long.

The true significance of the primes lies in the fact that all numbers in between can be made by multiplying primes together: 4 is 2 × 2, 6 is 2 × 3, 8 is 2 × 2 × 2, 9 is 3 × 3 and so on. So, by understanding the properties of the primes, we can understand other deep properties of numbers generally.

The ancient Greeks understood many aspects of primes. In perhaps the first great example of a watertight proof written 2,300 years ago, Euclid showed that there are infinitely many of them.

Euclid's prime proof

The Greek mathematician Euclid was active around 300 BCE in Alexandria in present-day Egypt. He is best known for his mathematical primer *The Elements*, which laid down the basic rules of geometry and much else – including a proof that the list of prime numbers has no end.

Suppose someone claims to have a complete list of primes. Multiply all the primes on the list together and then add 1 to this number. This new number is by definition not divisible by any of the primes on the list – you will always get remainder 1. So the new number is either another prime itself or divisible by a prime that is missing from the list. If you add this new prime to the list, repeating the trick will always show that any finite list is missing a prime.

Other questions about the prime numbers are tougher to answer, however. The Riemann hypothesis, for example, which concerns how the prime numbers are distributed along the number line, is one of the seven great 'millennium problems' for which any solution is worth $1 million (see Chapter 7).

Large primes and cryptography

There is no largest prime number, but that has not stopped mathematicians from competing over the years to discover ever larger ones (see Figure 4.1).

Since 1996 this search has been dominated by the Great Internet Mersenne Prime Search, or GIMPS, a distributed project that uses freely downloadable software to sift through numbers to see whether they are Mersenne primes. Mersenne primes are prime numbers of the form $2^p - 1$, where p is itself a prime number. That makes them relatively easy to find, although checking that they are not divisible by any prime beneath them is still a computationally intensive activity.

In 1999 GIMPS found the first one-million-digit prime number. The largest known prime number currently has over 22 million digits. Holding the record for the largest prime number may be all to do with the glory, but large prime numbers are practically very important. Internet banking, online shopping and digital authentication all use encryption, scrambling messages so that only the intended recipient knows how to de-scramble them – and that relies on prime numbers.

The idea is that the person looking to receive an encrypted message multiplies two big prime numbers together to create a new number that forms part of a public key, which they share with anyone wanting to send them a message. Anyone

Rank	Value	No of digits	Discovered in	Mersenne?
1	$2^{74207281}-1$	22338618	2016	Yes
2	$2^{57885161}-1$	17425170	2013	Yes
3	$2^{43112609}-1$	12978189	2008	Yes
4	$2^{42643801}-1$	12837064	2009	Yes
5	$2^{37156667}-1$	11185272	2008	Yes
6	$2^{32582657}-1$	9808358	2006	Yes
7	$10223 \cdot 2^{31172165}+1$	9383761	2016	
8	$2^{30402457}-1$	9152052	2005	Yes
9	$2^{25964951}-1$	7816230	2005	Yes
10	$2^{24036583}-1$	7235733	2004	Yes

FIGURE 4.1 The largest known prime numbers, as of August 2017

with the public key can encrypt messages. To decrypt them, however, turning them into something meaningful, requires knowledge of the two original prime numbers. Given a large enough number, working out the prime numbers that make it up is practically impossible, as the only way to do it is essentially to try all the different possibilities. This means that only the person who generated the public key can decrypt any messages, making the whole process secure.

The twin primes conjecture

As we climb to higher and higher numbers, the primes get more spread out on average. That squares with intuition and experience: the first few primes, 2, 3 and 5, are squashed up close together, but as you go up the number line there are more and more primes by which a number might potentially be divisible. The biggest two primes less than 100 are 89 and 97, eight apart.

But the details of the pattern are subtle – and unpredictable. Just after 100, we find the primes 101, 103, 107 and 109 all bunched up together. On average the primes get more spread out, but it seems that there are little pockets of prime activity.

Apart from 2 and 3, there cannot be any pairs of consecutive numbers that are both prime, because all even numbers apart from 2 are divisible by 2. Twin primes are pairs of primes that differ by 2, and there seem to be a lot of them. Examples are 3 and 5, or 41 and 43, or 107 and 109; the next twin prime years will be 2027 and 2029.

The twin primes conjecture predicts that, just as there are infinitely many prime numbers, there are infinitely many pairs of twin primes. It's hard to pin down when it was first proposed; it may go back as far as the ancient Greeks. There are good reasons to think it is true. But although since the mid-nineteenth century generations of notable number theorists have tackled the problem, there has been no conclusive proof – and until recently little sign of progress towards one.

That changed in April 2013, with a proof by a then unknown mathematician called Yitang Zhang working at the University of New Hampshire in the USA (see Figure 4.2). He showed that there are infinitely many pairs of primes where the gap is less than or equal to 70 million. That doesn't sound impressive when the number we are hoping for is just 2, but it was the first time that anyone had managed to get a finite bound – and 70 million is much smaller than infinity.

Against the stereotype of mathematicians doing their best work when they are young, Zhang was then in his late fifties, and had only finally managed to get an academic job after a spell working for the Subway restaurant chain following his PhD. Zhang's figure of 70 million was not the best that his approach would give, so others were able to tighten up the details of the

FIGURE 4.2 Yitang Zhang of the University of New Hampshire surprised the world in 2013 with a significant step forward in proving the notorious twin primes conjecture.

proof to do better. 'I just can't resist: there are infinitely many pairs of primes at most 59,470,640 apart' wrote Australian mathematician Scott Morrison on his blog at the end of May 2013.

Soon Fields Medallist Terence Tao started an online collaboration to tackle the problem more systematically. By the end of July 2013 the collaboration had used Zhang's method to show that there are infinitely many pairs of primes where the gap is less than or equal to just 4,680. In November 2013 James Maynard, who had just completed a PhD in analytic number theory at the University of Oxford, devised a simpler version of Zhang's method and sank the bound to just 600. By April 2014 the online collaboration had used the improved method to show that there are infinitely many pairs of primes with a gap that is less than or equal to 246.

Since then, there has been no progress. The approaches used so far all use a technique called sieve theory that traces its origins back to Eratosthenes of Cyrene, a Greek mathematician of the third century BCE. There is a well-known obstruction that stops mathematicians reaching a definitive solution by this route. A workaround is as yet unclear – and the twin primes conjecture, though much closer to resolution, remains frustratingly unproved.

The largest twin primes

As of August 2017, the largest pair of primes separated by just 2 are:

$$2996863034895 \times 2^{1290000} + 1$$

and

$$2996863034895 \times 2^{1290000} - 1$$

Are the primes random?

Whether a number is prime or not is pre-determined. But mathematicians have no way to predict which numbers are prime, so they tend to treat them as if they occur randomly. In 2016, however, Kannan Soundararajan and Robert Lemke Oliver of Stanford University in California confirmed that this is not quite right.

Apart from 2 and 5, all prime numbers end in 1, 3, 7 or 9 – they have to, or else they would be divisible by 2 or 5 – and each of the four endings is equally likely. But while searching through the primes, Soundararajan and Oliver noticed that primes ending in 1 were followed by another prime ending in 1 only 18.5 per cent of the time. If the primes were truly random, you'd expect that to be 25 per cent of the time – consecutive primes should not care about their neighbour's digits.

Similar patterns showed up for the other combinations of endings. They are also apparent in other bases, where numbers are counted in units other than 10s, meaning that the patterns are not a result of our number system but inherent to the primes. The patterns become more in line with randomness as you count higher – the pair have checked up to a few trillion – but still persists.

In fact, the result might not be such a surprise at all. In the early twentieth century G. H. Hardy and John Littlewood, two mathematicians who worked together at the University of Cambridge, came up with a way to estimate how often pairs, triples and larger grouping of primes will appear, known as the κ-tuple conjecture. The κ-tuple conjecture has not yet been proved, but it, too, suggests that the primes are not entirely randomly distributed – and this latest work is in line with that prediction.

Even so, for Soundararajan the finding was a salutary experience. 'It was very weird,' he says. 'It's like some painting you are very familiar with, and then suddenly you realize there is a figure in the painting you've never seen before.'

The Goldbach conjecture

The Goldbach conjecture predicts that every even whole number bigger than 2 can be written as a sum of two prime numbers – for example, $10 = 3 + 7$, and $78 = 31 + 47$. Notoriously, it has remained unproved ever since Christian Goldbach first proposed it in 1742.

In 2013, however, Harald Helfgott of the Ecole normale supérieure in Paris, France, proved a related problem: the odd Goldbach conjecture, which states that every odd number above 5 is the sum of three primes. A proof of

---

Goldbach's conjecture would also prove the odd version, since you can then take an even number formed of two primes and add 3 to it to get an odd number formed of three primes. But Helfgott's proof is unlikely to help mathematicians go in the other direction – so Goldbach's original problem remains unsolved.

5
π, φ, *e* and *i*

Beyond the intricacies of zero and infinity and the atomistic music of the primes, certain numbers fascinate because they are just there, mysteriously intrinsic to the structure of reality. In this chapter, we investigate four of these strange, essential intruders.

π: *the most famous ratio*

Every year on 14 March − 3/14 in the American way of writing dates − mathematics enthusiasts celebrate Pi Day. Pi, or π to use its usual Greek symbol, is the most ubiquitous mathematical ratio. Defined as the circumference of a circle divided by its diameter, its role in geometry means that it occurs in many mathematical formulae for calculating the surfaces, area and volume of shapes.

But its influence in mathematics and physics spreads much further. π is a central component, for example, of Fourier transforms, used to break down and analyse waveforms in electronics and elsewhere. It also pops up in Heisenberg's uncertainty principle in quantum mechanics, and to describe the geometry of space and time itself in Albert Einstein's equations of general relativity. In describing how things work in the universe, it's difficult to get round π.

The first few digits of π are 3.14159265, but notoriously it doesn't stop there. π is an irrational number, which means that its decimal representation goes on for ever. We don't need all those decimal places: NASA uses only around 15 digits of π in its calculations for sending rockets into space, and to get an atom-precise measurement of the universe you would need only around 40. Like the race to find the largest prime numbers (see Chapter 4), efforts to compute trillions of digits of π are mostly about the glory, or showing off computer power.

π's infinite length means that potentially every possible number you can think of is hidden somewhere in it − your date of birth, phone number, or even your bank details. Using a code that converts numbers into letters would let us find the Bible, the complete works of Shakespeare or, indeed, every book ever written, if we looked at enough digits. At least, that's the theory.

Normal, or not?

For π to encode everything, it would have to be a 'normal' irrational number, and we do not yet know if it is. If it is normal, the numbers 0 to 9 will appear equally often in its decimal representation. This means that any single-digit number occurs 10 per cent of the time, any two-digit number 1 per cent of the time, and so on.

The probabilities get vanishingly small when you start looking for the huge numbers of digits corresponding to Shakespeare's works, but if π is normal you would get there in the end. Many people are interested in the normality of π, even though proving it either way is unlikely to have much real-world impact.

The latest effort, which calculated over 22 trillion digits of π in November 2016, supports the idea that it is normal: each of the digits from nought to nine appeared 10 per cent of the time. But settling the normality of π for good cannot be done with calculations alone – it will require a mathematical proof. Twenty-two trillion digits might seem like a lot of good evidence, but compared with the infinitude of π, it's nothing.

Infinite π

In November 2016, after 105 days of round-the-clock computation, π enthusiast Peter Trueb's computer finally calculated 22,459,157,718,361 fully verified digits of π.

The efforts of Trueb, who is an R & D scientist by day, mean that an extra 9 trillion digits after the decimal point have been discovered, smashing the previous world record set back in 2013. It required a computer with 24 hard drives, each containing 6 terabytes of memory, to store the huge quantity of data produced with each step of the

---

process. To run the calculations, he used a computer program called γ-cruncher, which was developed by Alexander Yee and is available for free online.

Yee developed γ-cruncher as a hobby when he was in high school, and now works for a hedge fund in Chicago. The software uses what is known as the Chudnovsky algorithm for calculating π. It is quite a complicated mathematical formula, but what really makes γ-cruncher useful is its ability to perform calculations with trillions of digits. Yee likens the problem to trying to multiply two numbers that are a trillion digits long on a blackboard. It just would not work. So instead, he has introduced many fancy algorithms to streamline the calculations.

This is not the first time γ-cruncher has broken π world records, but previously Yee had himself played a part. This time it came as a complete surprise. The first he knew of it was when Trueb sent him an email saying that he had broken the world record.

The final file containing the 22 trillion digits of π is nearly 9 terabytes in size. If printed out, it would fill a library of several million books containing a thousand pages each.

π *everywhere you look …*

π's central role in geometry and mathematics means that it crops up in the oddest places:

- **… in the sky**
  The stars overhead inspired the ancient Greeks, but they probably never used them to calculate π. In 1994 Robert Matthews of the University of Aston in Birmingham, UK, combined astronomical data with

number theory to do just that. Matthews used the fact that for any large collection of random numbers, the probability that any two have no common factor is $6/\pi^2$. Numbers have a common factor if they are divisible by the same number, not including 1. For example, 4 and 15 have no common factors, but 12 and 15 have the common factor 3.

Matthews calculated the angular distance between the 100 brightest stars in the sky and turned them into 1 million pairs of random numbers. Around 61 per cent of them had no common factors, leading to a value for π of 3.12772, which is about 99.6 per cent correct.

- **... in winding rivers**
  Back on Earth, π controls the path of rivers from the Amazon to the Thames (see Figure 5.1). A river's meandering is described by its sinuosity – the length along its winding path divided by the distance from source to ocean as the crow flies. It turns out that the average river has a sinuosity of about 3.14.

- **... in books**
  π has inspired a particularly tricky form of creative 'constrained' writing called Pilish. These are poems – or 'piems' – where the number of letters of successive words is determined by π. One of the most ambitious piems is the *Cadaeic Cadenza* by Mike Keith. It begins with the lines: *One/A poem/A raven*, corresponding to 3.1415, and continues for 3835 word-digits. Keith has also written a 10,000-word book using the technique.

- **... in your front room**
  You can calculate π at home with some needles and a sheet of lined paper. Drop the needles on the paper

FIGURE 5.1 Divide the length of the average river by the distance as the crow flies from source to mouth, and the answer is π.

and calculate the percentage that cross a line where they fall. With enough attempts, the answer should be the needle length divided by the width between lines, all multiplied by 2/π.

This is known as Buffon's needle problem, after the French mathematician Georges-Louis Leclerc, Comte de Buffon, who first proposed it in 1733. The theory was put to the test in 1901 by Mario Lazzarini, a mathematician who dropped 3,408 needles to get a value of 3.1415929, correct to the first six decimal places. Subsequent examination of his results suggests that Lazzarini might have fiddled things, however, choosing numbers for the needle length and line width that gave the answer 355/113, a well-known approximation to π.

Interview: Michael Hartl and tau, π's nemesis

Michael Hartl has a PhD in physics from the California Institute of Technology, and is the author of the web development book Ruby on Rails Tutorial. *He says it's time to kill off* π, *since he believes that an alternative mathematical constant will do its job better.*

What's wrong with π?

Of course, π is not 'wrong' in the sense of being incorrect. It's just a confusing and unnatural choice for the circle constant. π is the circumference of a circle divided by its diameter, and this definition leads to annoying factors of 2. Try explaining to a 12-year-old why the angle for an eighth of a pizza – one slice – is π/4 instead of π/8.

So what should replace π?

In 2010 in my book *The Tau Manifesto*, I suggest using the Greek letter tau, which is equal to 2π, or 6.28318 … instead. Tau is the ratio of a circle's circumference to its radius, and this number occurs with astonishing frequency throughout mathematics.

If this idea is so fundamental, why haven't we made the change?

The paper that got the ball rolling on this, 'Pi is Wrong!' by mathematician Robert Palais, traces the history of π. It's only in the last 300 years that this convention was adopted, and I think it's just a mistake. It's one of those times in history when we chose the wrong convention.

Doesn't using tau ruin equations like the formula for circular area?

Quite the opposite. As I showed in *The Tau Manifesto*, using tau reveals underlying mathematical relationships that are obscured by using π. In particular, the famous formula for circular area is the manifesto's *coup de grâce*.

Has anyone successfully changed notation like this in the past?

In physics, there is an important quantity known as Planck's constant, h. As quantum mechanics developed, it became clear that h-bar was more important, which is equal to $h/2\pi$ – that's the factor of 2 that pops up everywhere! h-bar is now the standard notation, though both are used.

You are up against a formidable enemy, because π is a popular constant ...

It is: there are books written about it, and people care enough to memorize tens of thousands of its digits. Google has even changed its logo on Pi Day.

People celebrate π by eating pies on Pi Day. How can people celebrate 26 June, Tau Day?

If you think that the circular baked goods on Pi Day are tasty, just wait – Tau Day has twice as much pie!

ϕ: Fibonacci and the golden ratio

The Italian mathematician Leonardo of Pisa, better known as Fibonacci, has already been mentioned as the man who brought zero to Europe (see Chapter 2). But he is best known for the Fibonacci sequence, that list of numbers where the next digit is given by adding the previous two: 1, 1, 2, 3, 5, 8, 13, 21, 34, 55 and so on.

Fibonacci numbers crop up in nature. For example, the number of petals in a flower tends to be one of the Fibonacci numbers. This strange numerology can be traced to the dynamical behaviour of the cells at the tip of a growing shoot. The 'primordia' – tiny collections of cells from which the interesting features of plants develop – become arranged in patterns like interpenetrating spirals. The mathematics of such patterns leads inevitably to Fibonacci numbers.

The Fibonacci sequence is an example of a complete sequence. Any positive whole number can be expressed as a sum of terms in the sequence, using no term more than once: 4 is 1 + 1 + 2, for example. But work out the ratio between consecutive terms and, as the terms get larger, you'll quickly find yourself getting nearer and nearer to a specific number, whose first few digits are 1.618.

This mysterious number is most famously called the golden ratio, represented by the Greek letter phi, ϕ, and it crops up a lot. Try drawing a diagonal line connecting two corners of a regular pentagon. Divide the length of that line by the length of the pentagon's sides and out it pops. Something similar is possible with an equilateral triangle. In general, if you have a small number, A, and a larger one, B, and set the numbers so that the ratio of B to A is the same as A+B to B, then that ratio is always the golden ratio (see Figure 5.2).

Search for the golden ratio online and you will be inundated with claims of how ancient Greek architecture and the human face exhibit this ratio, and that people find it immensely aesthetically pleasing. But the evidence here is murky. The human body has thousands of different proportions; some of them seem to be close to the golden ratio for some people. Ancient Greek mathematicians and architects were definitely aware of the golden ratio, but the ruins left over are a bit crumbly and

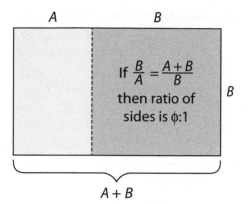

FIGURE 5.2 Shapes whose proportions conform to the golden ratio, ϕ, are supposedly particularly aesthetically pleasing.

have many varying proportions; look hard enough and you'll find the golden ratio if you really want to.

A similar problem plagues studies that ask people to look at some artworks incorporating the ratio and others that do not, to rate how aesthetically pleasing they are. It is unclear whether that judgement is really based on the ratio and, even if it is, whether the association is learned or innate. It is a beautiful number all the same.

e: exponents and logarithms

In 2004 Google announced that it was aiming to raise $2,718,281,828 from the first sale of its shares. The exactness of the figure left many perplexed, but mathematicians nodded knowingly. The figure is one of the most important numbers in mathematics: the first ten digits of Euler's number, or e.

Along with π, e has transformed our understanding even of the concept of number. Both numbers exist in their own right and crop up throughout the natural world. The number e plays a key part in describing how things reproduce or grow – money and populations, for instance – and also decline. It occurs, for example, in the equation characterizing radioactive decay.

---

e and compound interest

Euler's number plays a pivotal role in how things grow, as the Swiss mathematician Jacob Bernoulli first showed.

Put £1 in the bank. If the yearly interest were 100 per cent – if only – a year later you will have £2. That's simple enough. But what if the interest were determined more regularly? Say, for example, your bank calculates the interest after six months and gives you 50 per cent to take your £1 to £1.50. Then, at the end of the year, you get another 50 per cent, making £2.25: a nice little improvement.

Carry on with the same logic and if the interest were compounded monthly, you would end up with £2.61. If it were daily, you would get £2.71. But, no matter how fast it is compounded, there is an amount you will never exceed – £2.718281828..., or *e*.

The number first surfaced in 1618, through the work of British mathematicians John Napier and William Oughtred on 'slide rules', convenient devices for multiplying large numbers in the days before calculators. In 1683 Swiss mathematician Jacob Bernoulli rediscovered *e* by studying how bank accounts grow as interest is added year after year (see '*e* and compound interest' above). But it was the work of Swiss genius Leonhard Euler in the eighteenth century that really placed *e* at the centre of the mathematical universe.

The number's value may be 2.718281828..., but mathematical definitions of *e* are more slippery. One is that it emerges as the result of the expression $(1 + 1/n)^n$ as n tends to infinity. Euler showed how the true significance of *e* lies in its connection to the mathematical operation called exponentiation. Like addition, subtraction, multiplication and division, exponentiation is a fundamental way to combine numbers. It is written as a^b, where a is the base and b the exponent, and is easy to define for integers: for example, 4^3 is 4 multiplied by itself 3 times: $4 \times 4 \times 4 = 64$. But for powers that are not whole numbers, this definition of exponentiation does not obviously work: what might it mean to multiply something by itself π times, for example?

Euler solved this problem by finding a way to define e^x where x doesn't have to be an integer, and then showed how to write every number a^b as e^x, providing an easy formula for finding x in terms of a and b. The fact that exponentiation can be extended to all numbers is one of the cornerstones of mathematics and revolves around *e*.

The inverse operation of exponentiation is called taking the logarithm. If 4^3 is 64, then the base-4 logarithm of 64 is 3; similarly, the base-10 logarithm of 100 is 2 because $100 = 10^2$. In situations where things are growing exponentially, logarithms provide a more manageable way to calculate things. The natural appearance of *e* in the formula for exponentiation means that logarithms based on *e* are particularly easy to deal with – so they are known as 'natural' logarithms.

Euler and his identity

Born in Basel in Switzerland in 1707, Leonhard Euler was a polymath who wrote books on, among other things, planetary orbits, ballistics, shipbuilding, navigation and calculus. But he is most remembered as a pioneer of mathematical analysis whose insights have set the direction of the subject ever since.

Most strikingly, he proved the formula $e^{i\pi} + 1 = 0$. This expression has become central to our understanding of number and exponentiation, and is celebrated for the beautiful way it unites five fundamental constants of mathematics: zero, one, e, π and i, the square root of -1.

After demonstrating a proof of this equation in a lecture, the nineteenth-century American mathematician Benjamin Peirce is reputed to have told the audience: 'Gentlemen ... it is absolutely paradoxical; we cannot understand it, and we don't know what it means. But we have proved it, and therefore we know it is the truth.' The Nobel prize-winning physicist Richard Feynman described it as 'the most remarkable formula in mathematics'.

Transcendental numbers

e and π are both examples of transcendental numbers, a type of number whose baffling complexity is the antithesis of the everyday integers 0, 1, 2, 3, 4 and so on.

A transcendental number cannot be written as a fraction – it is irrational. But it is also entirely unrelated to the integers by any sequence of ordinary arithmetical operations. You can multiply a transcendental number by itself as many times as you wish, combine the numbers that result and divide and multiply by integers any way you want, but you will never arrive back in the familiar territory of the integers.

For centuries, no one even suspected that such strange objects might exist. The ancient Greeks believed that all numbers could be derived from the integers by simple division. According to legend, in around 500 BCE, when Hippasus of

Metapontum proved that some numbers, such as the square root of 2, couldn't be written as fractions involving integers, his fellow Pythagoreans were so outraged that they had him drowned for heresy.

But irrational numbers like the square root of 2 are tame compared with transcendentals. By definition, the square root of 2 times itself equals 2, so we get back to an integer after just one step. The first rock-solid transcendental number was discovered in 1844 by the French mathematician Joseph Liouville, although the idea has its roots in earlier work by Euler.

Transcendental numbers might have been regarded as a curiosity had not another French mathematician, Charles Hermite, proved in 1873 that *e* is transcendental. Within ten years, π had been added to the list. Then the set theorist Georg Cantor delivered a thunderbolt. Far from being a handful of exotic, if significant, anomalies, in fact almost all numbers are transcendental: they infinitely outnumber the non-transcendentals.

The consequences of Cantor's work are profound. It means that the range of numbers that human brains and also computers are equipped to handle – those easily derived from the integers – are just an infinitesimal sliver of the numerical universe. Swarming around the familiar integers and fractions is an infinitely larger collection of transcendental numbers – the 'dark matter' of the numerical universe.

i: *the imaginary number*

Transcendental numbers might be enough to send you to another plane, but what about numbers that are entirely imaginary? Basic rules of mathematics say that two positive numbers multiply to give a positive, and two negative numbers also multiply to give a positive. So what number could you multiply by itself to give -1? Answer: an imaginary number.

Imaginary numbers have been lurking in mathematics since the sixteenth century at least. They popped up as geometers were investigating solutions to equations such as those with an x^2 or x^3 term in them, some of which seemed to involve the square roots of negative numbers. René Descartes first called these numbers 'imaginary' in 1637. He was being derogatory. But they, and his description, persisted. In the eighteenth century they came to be represented as multiples of i, the square root of -1.

i is not a 'real' number you can count and measure things with. You cannot work out whether it is divisible by 2, or less than 10. Imaginary numbers do not fit on the regular number line, so mathematicians put them on a second, independent line, with the two intersecting at zero (see Figure 5.3). The lines can be treated as axes, making 'complex' numbers – consisting of a real and an imaginary part – handy for representing things that change in two dimensions. In geometry they appear in trigonometric equations, and in physics they provide a neat way to describe rotations and oscillations. Electrical engineers use them routinely in designing alternating-current circuits, and they are useful for describing light and sound waves, too. They have also become essential tools in microchip design and digital compression algorithms for transporting and reproducing pictures and music.

More fundamentally, they are used in wave functions, the mathematical descriptions of particles in quantum mechanics, where, fittingly, they encapsulate the idea that things can be in two places or states at once. These multiple personalities are depicted by a series of complex numbers that describe the probability that a quantum particle has a particular property such as location or momentum. Whereas

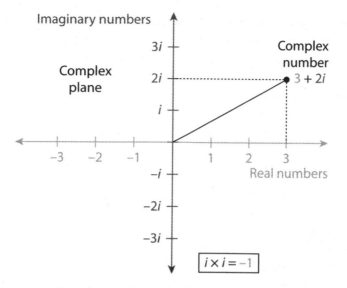

FIGURE 5.3 Complex numbers, which contain both real and imaginary elements based on the square root of –1, known as *i*, lie somewhere on a 2D 'complex plane'.

alternative real-number descriptions for something like a light wave in the classical world are readily available, purely real mathematics simply does not supply the tools required to paint a picture of the fuzzy quantum world.

And the truth is that 'real' and 'imaginary' numbers are both abstract concepts anyway. We might be more familiar with 5 than 5*i*, but neither can be found on its own in the real world. This gives mathematicians a certain creative licence. In 1843 the Irish mathematician William Hamilton invented additional solutions for the square root of –1 he called *j* and *k*. The 4D numbers he built on this basis, known as quaternions, are used to encode 3D rotations in some computer games.

If you follow the same mathematical logic, there is no reason to stop there. Today, octonions add an extra seven dimensions of imaginary numbers and the rarely used sedenions give the option of extending the total to 15. Up there, it's a world of pure imagination.

6

Probability, randomness and statistics

We live in an uncertain world, but here, too, numbers can help us make sense of things. If we can make sense of the numbers, that is. The world of probability and statistics is full of counter-intuitive results that can lead the unwary astray.

How to think about probability

The basic idea of probability is simple enough. It is the measure of how likely or unlikely something it is to happen, and is assigned a value from 0 to 1, where 0 means it never will happen and 1 means it is certain to happen. Alternatively, probability values are often expressed as a percentage between 0 and 100 per cent. But beyond those simple facts, probability is one of those things we all get wrong ... deeply wrong. The good news is that we are not the only ones. Even pure mathematicians claim that probability has many unreasonable answers.

Take the classic problem of a class of 25 schoolchildren. How likely is it that two of them share the same birthday? The intuitive, commonsense answer is that it is not implausible, but quite unlikely. You would be wrong.

Before revealing the answer, let's look at the celebrated Monty Hall problem, named after the former host of US television game show *Let's Make a Deal*. You're playing a game in which there are three doors, one hiding a car, two of them hiding goats. You choose one door; the host of the game then opens another, revealing a goat. Assuming you'd rather win a car than a goat, should you stick with your choice or swap? The naive answer is it doesn't matter: you now have a 50–50 chance of striking lucky with your original door. This would also be wrong (see Figure 6.1).

But if probability makes even experts grumble, how do we get it right? Simple, says mathematician Ian Stewart of the University of Warwick: do things the hard way. That means switching off your intuition, thinking carefully about how the problem is posed and do your sums diligently.

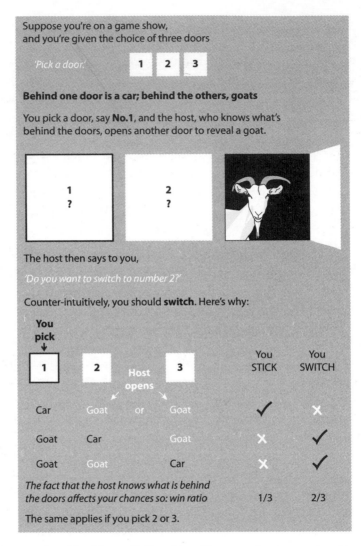

Suppose you're on a game show,
and you're given the choice of three doors

'Pick a door.'   **1** **2** **3**

Behind one door is a car; behind the others, goats

You pick a door, say **No.1**, and the host, who knows what's
behind the doors, opens another door to reveal a goat.

The host then says to you,

'Do you want to switch to number 2?'

Counter-intuitively, you should **switch**. Here's why:

You pick ↓ 1	2	Host opens	3	You STICK	You SWITCH
Car	Goat	or	Goat	✓	✗
Goat	Car		Goat	✗	✓
Goat	Goat		Car	✗	✓

The fact that the host knows what is behind the doors affects your chances so: win ratio   1/3   2/3

The same applies if you pick 2 or 3.

FIGURE 6.1 The Monty Hall problem illustrates the counter-intuitive nature of some results in probability.

With the birthday problem, the starting point is to realize that you are not interested in individual schoolchildren, but pairs. In a class of 25, there are 300 pairs to consider and in most years 365 days on which each might share a birthday. Factor all that in, and you end up crunching some truly astronomical numbers to arrive at the answer, which is actually just under 57 per cent. In other words, it is more likely than not that two children share a birthday: a major defeat for our intuition.

With the Monty Hall problem, meanwhile, the chance you chose the right door in the first place is 1/3 — and that doesn't change whatever happens afterwards. There's also a 2/3 chance that the car is behind one of the other two doors. And since the host has revealed a goat behind one of them, that 2/3 probability now applies to just the other unopened door — so you are better off swapping.

There are a few caveats: if the host is so devious as only to open a door if you chose the right one in the first place, you'd be mad to swap. Ditto if you want a goat rather than the car. That illustrates another important rule in thinking about probability, says John Haigh, a mathematician at the University of Sussex in Brighton, UK. It is very important to know your assumptions. Very subtle changes can change the outcome.

All this is very well when the boundaries of the problem are clear and the possible outcomes quantifiable. Toss a fair coin and you know you have a 50 per cent chance of heads — because you can repeat the exercise over and over again if necessary.

But what about a 50 per cent chance of rain today, or of a horse with even odds winning a race? No amount of expert advice can help us assess the true worth of such 'subjective' probabilities, which are fluid and often based on inscrutable expertise or complex modelling of an unpredictable world. And that exposes a truth about probability that is often overlooked: there is no one agreed way of going about calculating it.

Frequentist and Bayesian probability

We are in a bar, and agree to toss a coin for the next round. Heads, I pay; tails, the drinks are on you. What are your chances of a free pint? Most people – sober ones, at least – would agree: evens.

Then I flip the coin and catch it, but hide in it the palm of my hand. What's your probability of free beer now? Broadly speaking, there are two answers: (1) it is still 50 per cent, until you have reason to think otherwise; (2) assigning a probability to an event that has already happened is nonsense. Which answer you incline towards reveals where you stand in a 250-year-old debate on the nature of probability and statistics: the spat between frequentist and Bayesian statistics.

Drawing conclusions without all the facts is the bread-and-butter of statistics. How many people in a country support the legalization of cannabis? You cannot ask all of them. Is a run of hotter summers consistent with natural variability, or a trend? There is no way to look into the future to say definitively.

Answers to such questions generally come with a probability attached. But that single number often masks a crucial distinction between two different sorts of uncertainty: stuff we do not know and stuff we cannot know. Cannot-know uncertainty results from real-world processes whose outcomes appear random: how a die rolls, where a roulette wheel stops, when exactly an atom in a radioactive sample will decay. This is the world of frequentist probability: if you roll enough dice or observe enough atoms decaying, you can get a reasonable idea of the probability of different outcomes.

Don't-know uncertainty is more complicated. Here, individual ignorance, not universal randomness, is at play. The sex of a newly conceived baby or the winner of a future horse race are examples – don't-know uncertainties are those beloved of the bookies.

A strict frequentist has no truck with don't-know uncertainty or any probability measure that cannot be derived from repeatable experiments, random number generators, surveys of a random population sample and the like. A Bayesian, meanwhile, is unafraid to use other 'priors' – knowledge gleaned about the form of a horse from past races, for example – to fill in the gaps.

The coin-in-the-pub example shows where these two views diverge. Before the coin flip, frequentist and Bayesian probabilities line up: 50 per cent. After the flip the source of uncertainty changes from intrinsic randomness to personal ignorance. If you were inclined to Bayesian ways of working, you might quote a probability figure of 50 per cent – or perhaps a telltale flicker of a victorious smile on my face might persuade you to downgrade your chances. Bayesian advocates try to answer questions by bringing all the relevant evidence to bear on it, even when the contribution of some of that evidence depends on subjective judgements.

The Bayesian controversy

Bayesianism takes its name from the English mathematician and Presbyterian minister Thomas Bayes (1702–61) (see Figure 6.2). In an essay published in 1763, two years after his death, he set out a new approach to a fundamental puzzle: how to work backwards from observations to hidden causes when your information is incomplete.

Imagine you have a box of a dozen doughnuts, half cream, half jam-filled. It is relatively straightforward to calculate the probability of taking out five jam doughnuts in a row. But the backwards problem, working out the probable contents of an unknown box when you have just taken out five jam doughnuts, is trickier. Bayes's innovation was to provide the seed of a mathematical framework that allowed you to start with a guess (perhaps you have bought boxes of doughnuts from that store before) and refine it as further data came to light.

In the late eighteenth and early nineteenth centuries Bayesian-style methods helped tame a range of inscrutable problems, from estimating the mass of Jupiter to calculating the number of boys born worldwide for every girl. But it gradually fell out of favour, victim of a dawning era of big data. Everything from improved astronomical observations to newly published statistical tables of mortality, disease and crime conveyed a reassuring air of objectivity. By contrast, Bayes's methods of educated guesswork seem hopelessly old-fashioned and rather unscientific. Frequentist probability, with its emphasis on dispassionate number crunching of the results of randomized experiments, came increasingly into vogue.

The advent of quantum theory in the early twentieth century, which re-expressed reality in the language of frequentist probability, provided a further spur to that development. The two strands of thought in statistics gradually drifted further apart. Adherents ended up submitting work to their own journals, attending their own conferences and even forming their own university departments.

Emotions often ran high. The author Sharon Bertsch McGrayne recalls that when she started researching her book on the history of Bayesian ideas, *The Theory That Would Not Die*, one frequentist-leaning statistician berated her down the phone for attempting to legitimize Bayesianism. In return, some Bayesians developed a sort of persecution complex and others a kind of religious zealotry.

This is more than an esoteric problem. Larry Wasserman of Carnegie Mellon University in Pittsburgh, Pennsylvania, argues that the frequentist–Bayesian debate affects everybody's lives. A drugs company testing a new drug can come to very different conclusions according to which method it uses to analyse its results. Likewise, a jury might reach a different decision after hearing evidence presented in frequentist and Bayesian terms.

FIGURE 6.2 This portrait is claimed to be of Thomas Bayes, the originator of Bayesian probability, although it is uncertain whether it is of him as no other portrait survives.

Horses for courses

Both frequentist and Bayesian approaches have their strengths and weaknesses. Where data points are scant and there is little chance of repeating an experiment, Bayesian methods can excel in squeezing out information. A supernova explosion in a nearby galaxy, the Large Magellanic Cloud, seen in 1987, provided a chance to test long-held theories about the flux of neutrinos from such an event – but detectors picked up only 24 of these slippery particles. Without data, frequentist methods fell down, but the flexible, information-borrowing Bayesian approach provided an ideal way to assess the merits of different competing theories.

It helped that well-grounded theories provided good priors to start that analysis. Where these do not exist, a Bayesian analysis can easily be a case of 'garbage in, garbage out'. It is one reason why courts of law have been wary to adopt Bayesian methods, even though on the face of it they are an ideal way to synthesize messy evidence from many sources. In a 1993 New Jersey paternity case that used Bayesian statistics, the court decided that jurors should use their own individual priors for the likelihood of the defendant having fathered the child, even though this would give each juror a different final statistical estimate of guilt.

Finding good priors can also demand an impossible depth of knowledge. Researchers searching for a cause for Alzheimer's disease, for instance, might test 5,000 genes. Bayesian methods would mean providing 5,000 priors for the likely contribution of each gene, plus another 25 million if they wanted to look for pairs of genes working together. Generating a reasonable prior from such a high-dimensional problem would be nigh-on impossible.

To be fair, without any background information, standard frequentist methods of sifting through many tiny genetic effects would have a hard time letting the truly important genes and combinations of genes rise to the top of the pile. But this is perhaps a problem more easily dealt with than conjuring up 25 million coherent Bayesian guesses.

Frequentism in general works well where plentiful data should speak in the most objective way possible. One high-profile example is the search for the Higgs boson, completed in 2012 at the CERN particle physics laboratory near Geneva, Switzerland. The analysis teams concluded that if, in fact, there were no Higgs boson, then a pattern of data as surprising as, or more surprising than, what was observed would be expected in only one in 3.5 million hypothetical repeated trials. That is so unlikely that the team felt comfortable rejecting the idea of a universe without a Higgs boson.

That wording may seem convoluted, and this highlights fre-quentism's main weakness: the way it ties itself in knots through its disdain for all don't-know uncertainties. The Higgs boson either exists or it doesn't, and any inability to say one way or the other is purely down to lack of information. A strict fre-quentist cannot actually make a direct statement of the prob-ability of its existing or not – as, indeed, the CERN researchers were careful not to (although certain sections of the media and others were freer).

Head-to-head comparisons point to the confusions that can arise, as was the case in the 1990s with a controversial clinical trial of two heart-attack drugs, streptokinase and tissue plasmi-nogen activator. The first, frequentist analysis gave a 'p value' of 0.001 to a study seeming to show that more patients survived after the newer, more expensive tissue plasminogen activator therapy. This equates to saying that if the two drugs had the same mortality rate, then data at least as extreme as the observed rates would occur only once in every 1,000 repeated trials.

What does that mean? Not that the researchers were 99.9 per cent certain the new drug was superior – although, again, it is often interpreted that way. When other researchers conducted a Bayesian re-analysis using the results of previous clinical trials as a prior, they found a direct probability of the new drug being superior of only about 17 per cent. The real value of Bayesianism is that it directly addresses how likely the question of interest is to be true, says statistician David Spiegelhalter of the University of Cambridge. 'Who wouldn't want to talk about that?'

Sometimes, Bayesian and frequentist ideas can be blended to create something new. In large genomics studies, a Bayesian analysis might exploit the fact that a study testing the effects of 2,000 genes is almost like 2,000 parallel experiments, and cross-fertilize the analyses using the results from some to establish priors for others to hone the conclusions of a frequentist analysis.

Randomness

Probability is used to allow us to understand random phenomena, where the outcome of any one event is unknowable. Again, the coin flip is a classic example. Whether any single flip comes up heads or tails is random, and nothing is going to tell us which it will be. But by observing many coin flips over time, we can begin to scope out the problem: with an unbiased coin it becomes apparent that heads and tails both occur 50 per cent of the time.

Many natural phenomena are random; the problem is that our brains are not. We are wired to spot and generate patterns. That is useful when it's all about seeing predators on the savannah before they see you, but it handicaps us when we are dealing with random phenomena. It also makes it very difficult for us to produce randomness ourselves. True randomness is a useful thing to have, in devising secure keys for encryption, for example, and in many areas of computing, scientific modelling and design. If we want true randomness, we have to have a way of generating random numbers.

We might use a coin flip to produce a random string of 1s and 0s, for example, but this is a wearisome process and systematic effects – such as a slight loading of the coin – may render the results not truly random. The first dice for divination and gaming were six-sided bones from the heels of sheep, with numbers carved into the faces. The shape made some numbers more likely to appear than others, giving a decisive advantage to those who understood its properties. Suspicion about the reliability of randomness generators remains with modern equivalents like casino dice, roulette wheels and lottery balls.

Modern random number generators use a set algorithm to 'seed' a random-looking output from a smaller, unpredictable input: use the date and time to determine which random digits to extract from a random number string such as π, say, and work from there. The problem is that such 'pseudo-random' numbers are limited by the input and tend to repeat non-randomly after a certain time, in a way that is guessable if you see enough of them.

An alternative is to hook a computer up to some source of physical, 'true' randomness. In the 1950s the UK Post Office wanted a way to generate industrial quantities of random numbers to pick the winners of its Premium Bonds lottery. The job fell to the designers of the pioneering Colossus computer, developed to crack Nazi Germany's wartime codes. They created ERNIE, the Electronic Random Number Indicator Equipment, which harnessed the chaotic trajectories of electrons passing through neon tubes to produce a randomly timed series of electronic pulses that seeded a random number. ERNIE, now in its fourth iteration, now relies on thermal noise from transistors to generate randomness. Many modern computing applications use a similar source, collected using on-chip generating units.

Two problems remain. First, with enough computing power anyone can, in theory, reconstruct the processes of classical physics that produced or seeded the random numbers. Second, and more practically, random number generators based solely on physical processes are often unable to produce random bits fast enough.

Many systems, such as the Unix-based platforms used by Apple, get round the first problem by combining the output from on-chip randomness generators with the contents of an 'entropy pool', filled with other random contributions. This could be anything from thermal noise in devices connected to the computer to the random timings of the user's keyboard strokes. The components are then combined using a 'hash function' to generate a single random number. Hash functions are the mathematical equivalent of stirring ink into water: there is no known way to work out what the set of inputs was, given the number the function throws out. That doesn't mean that there couldn't be in the future – and there is still the speed problem. The workaround is generally to use a physical random number generator only as a seed for a program that generates a more abundant flow.

But a program means rules. Its output cannot be truly random, and is in principle guessable by someone who knows the code. The precise nature of the methods random number generator programs use is proprietary, but in 2013 security analysts raised concerns that the US National Security Agency knew the internal workings of one such generator, called Dual_EC_DRBG, potentially allowing them to break encryptions that relied on it. If you are just playing online games, that's not a big problem. But when making multibillion-dollar financial transactions or encrypting sensitive documents, a suspicion that you are being watched is a bigger deal.

Quantum randomness

Some researchers think that we shall never have an uncrackable source of randomness as long as we rely on the classical world. Here, randomness is not intrinsic but down to who has what information. For safer encryption, we must turn to quantum physics, where things truly do seem random.

Instead of a coin toss, you might ask whether a photon hitting a half-silvered mirror passed through it or was reflected. Instead of rolling a die, you might present an electron with a choice of six circuits to pass through.

Cryptographic systems that exploit the vagaries of quantum theory for more secure communication do exist. But they are not the last word in security. Extracting quantum randomness always involves someone making non-random choices about equipment, measurements and suchlike. The less-than-perfect efficiency of photon detectors used in some methods could also provide a back door through which non-randomness can slip in.

One way out that is still under investigation might be to amplify quantum randomness, so that you always have more of it than anyone can hack. Ways exist in theory to turn n random bits into 2^n bits of pure randomness, and also to launder bits to remove any correlation with the device that first made them. The question that remains is how to put these methods into practice.

Lies, damn lies and ...

In 1946 the British epidemiologist Austin Bradford Hill ran the first medical clinical trials in which participants were randomly assigned to two groups, one of which received treatment and one of which did not. One of these trials tested the effectiveness of the antibiotic streptomycin to treat tuberculosis, a condition that Bradford Hill himself had developed while serving in the First World War.

After just six months, the results were so convincing that they led to streptomycin being adopted as the standard treatment. In 1950, together with Richard Doll, Bradford Hill similarly pioneered the use of statistical methods to provide the first convincing evidence of a causative link between smoking and lung cancer.

From trials of the latest wonder drug to the discovery of the Higgs boson, breakthroughs that advance human knowledge are now seldom made without someone somewhere applying statistical reasoning. And as those bits of knowledge filter down to the rest of us, we are increasingly expected to make decisions on the basis of statistics.

Human health is a particularly fraught area that attracts more than its share of lurid statistically based headlines: that some common lifestyle factor increases the likelihood of developing a particular form of cancer by 50 per cent, or that some miracle cure reduces it by a similar amount. Even if these statistics are right, misinterpretations can cloud our judgement of them. For example, we often read of a test being so-and-so per cent accurate, without realizing that this is a meaningless figure unless we also know the test's false-positive rate (see Figure 6.3). And there are many other ways in which incomplete or ill-conceived statistics can mislead, as the following examples show.

You've just been diagnosed with a rare condition that afflicts 1 in 10,000. The test is 99 per cent certain. Hope or despair?

■ True positive   ▫ False positive

In a population of 10,000, on average one person will have the disease – and they will also test positive.

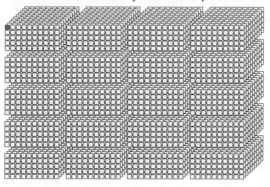

If the test is only 99 per cent accurate, 1 per cent of the remaining, healthy population will test positive too.

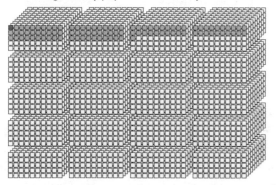

So if you test positive, all other things being equal, there's a chance of over 99 per cent you don't have the disease – hope.

FIGURE 6.3 False positives in screening tests may lead to a false assessment of how likely you are to have a disease.

Ratio bias

What would worry you more: being told that cancer kills 25 people out of 100, or that it kills 250 people out of 1000? A silly question, you might say; both statements mean that a quarter of people die of cancer. Yet, in unguarded moments, we are prone to see the second statement as equating to a bigger risk.

In one study of this 'ratio bias' effect, people rated cancer as riskier when told that it 'kills 1286 people out of 10,000' than when told it 'kills 24.14 people out of 100', even though the second statement equates to almost double the risk. Similarly, 100 people dying from a particular form of cancer every day can be perceived as a lesser risk than 36,500 dying from the same disease each year, although the two are equivalent statements.

When confronted with questions of risk, we therefore need to look carefully at the way the numbers are presented. And if we are comparing risks, we need to make sure that they are divided by the same number.

Relative vs absolute risk

The discovery in 2007 that a daily bacon sandwich raises the likelihood of bowel cancer by 20 per cent was deliciously encapsulated by the British tabloid newspaper *The Sun* in the headline 'Careless pork costs lives'. Such assertions may or may not be valid, but hidden within them is a more insidious source of confusion. They tend to quote relative risks: how much more likely you are to get ill when indulging in the supposedly dangerous substance or activity compared with not indulging. But they tell you nothing about what that increase in risk amounts to in absolute terms, so there is no way of telling whether it is something worth being concerned about.

For an average person, the chance of getting bowel cancer at some point in their life is around 5 per cent. So a 20 per cent relative increase in bowel cancer risk translates to an absolute increase in risk from 5 per cent to 6 per cent – just 1 per cent. That's big enough not to ignore, but less of a deterrent to those who like their daily bacon sandwich (see Figure 6.4).

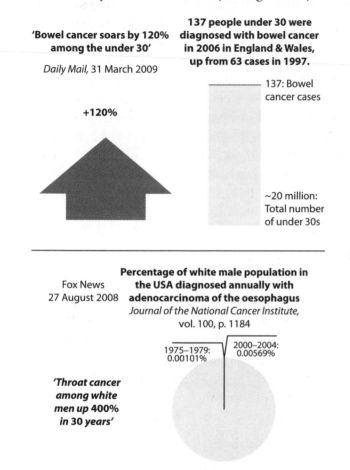

FIGURE 6.4 Different ways of presenting the same data can greatly influence our perception of risk.

Causation vs correlation

In 2010 a study entitled 'Television viewing time and mortality' grabbed the headlines. It had asked 8800 people about their health, lifestyle and television watching behaviour, and then followed them over the next six years. During that time, 284 of them died. Among people who spent more than four hours a day in front of the TV, it found, the risk of their dying within the period of the study was 46 per cent higher than among those who watched less than two hours a day.

The sorts of headlines generated – 'TV kills, claim scientists' – were predictable. You will already have noticed that it is a relative, not an absolute, risk being quoted. But this is also a case where two variables moving in tandem (correlation) do not necessarily mean that the change in one is responsible for change in the other (causation). More work would be needed to establish whether there is anything intrinsic to watching television that makes people more likely to die. In the meantime, there are other potential, and perhaps more likely, explanations. For example, people with certain underlying health problems sit or lie still for long periods, possibly in front of the TV, and these problems might also be associated with a raised risk of early death.

Before assigning cause and effect, it is essential to read between the lines. Austen Bradford Hill identified the crucial question in his pioneering statistical studies: is there any other way of explaining the set of facts before us; and is any such explanation equally, or more, likely than cause and effect? The answer needs to be a resounding no.

Statistical significance

'Over 80 per cent of women say that this shampoo leaves their hair healthier and shinier.' Such claims are common in advertising for all manner of consumer products. What they might not tell you is that only five women tested the shampoo. And of the four who certified its miraculous effect, one or two probably ended up with nicer hair purely by chance, or simply imagined the results.

Similar caveats apply to the effectiveness of medical treatments. Curing six out of ten patients is promising. Curing 300 out of 500 is the same success rate, but far more convincing. The sample size in a test is crucial in deciding whether any apparent improvement could have happened by chance alone, says University of Cambridge statistician David Spiegelhalter.

The standard procedure for clinical trials is the one established by Austen Bradford Hill all those years ago: new medical treatments are tested in randomized controlled trials (RCTs), in which volunteers are randomly allocated to a study group that receives the new treatment or a control group that receives a placebo or an existing treatment. The smaller the expected effect of a drug, the more people you need to test it on. Even then, it is possible that a useless treatment will register the effect you are after as a result of chance alone – which is one reason why drug licensing authorities do not usually consider a single study sufficient evidence to approve a new drug.

The next time you hear of public acclaim for a miracle cure or wonder shampoo, ask three questions. How many people was it tested on? Was it tested in an RCT? And was the result confirmed by a second, independent test?

Survival vs mortality rates

In his campaign to win the 2008 Republican presidential nomination, former New York City mayor Rudy Giuliani quoted the chance of a man in the US surviving prostate cancer – a disease he had developed – as 82 per cent. He then compared this with the 44 per cent chance of survival under the UK's taxpayer-funded National Health Service.

If right, that would surely be a damning indictment of the deadly inadequacy of socialized medicine, as Giuliani was seeking to emphasize. And his figures were right. However, they were also misleading. Giuliani was quoting five-year survival rates, the number of people diagnosed with a disease in a given year who are still alive five years later. But while prostate cancer in the US is generally diagnosed through screening, in the UK it is diagnosed on the basis of symptoms. Screening tends to pick up the disease earlier, leading to one source of bias in the comparison.

Suppose that of a group of men with prostate cancer all die at the age of 70. If the men do not develop symptoms until they are 67 or later, the five-year survival rate based on a symptoms approach is 0 per cent. Suppose, on the other hand, that screening had picked up the cancer in all of these men at age 64. Despite the outcome being the same – all the men dying at 70 – the five-year survival rate in this case is 100 per cent.

Yes, you might say, but earlier diagnosis through screening increases the chance that corrective measures can be taken. But screening is not 100 per cent accurate. First there are false positives, in which the test incorrectly flags a healthy person as having cancer. Prostate screening also picks up non-progressive cancers, which will never lead to symptoms, let alone death. The exact extent of this over-diagnosis is unclear, but a rough estimate is that 48 per cent of men diagnosed in this way don't have a progressive form of the cancer.

False diagnosis and over-diagnosis both result in unnecessary treatment, and, potentially, significant harm – in the case of prostate cancer, men left impotent and incontinent. Over-diagnosis also inflates the five-year survival rate by including men who would not have died of prostate cancer anyway.

A far better comparative measure of outcomes than the survival rate is the mortality rate, the proportion of people in the whole population who die from a condition in a given year. Figures from the period 2003 to 2007 published by the US National Cancer Institute indicate an age-adjusted mortality from prostate cancer of 24.7 per 100,000. Similar figures from Cancer Research UK for 2008 point to a mortality of 23.9 per 100,000. In statistical terms, that is a dead heat. Higher survival does not necessarily mean fewer deaths – and any time you see survival rates quoted to make a point, it's worth engaging in some statistical scepticism.

Interview: The man who gave us risk intelligence

In 2011 Dylan Evans, who has a PhD in philosophy from the London School of Economics, founded risk intelligence company Projection Point. Humans are no good at assessing probabilities but, against the odds, Evans has tracked down the handful of people who rate as geniuses on the intelligence scale he calls risk quotient.

Most people probably haven't heard of risk intelligence. What is it?

It is the ability to estimate probabilities accurately. It's about having the right amount of certainty to make educated guesses. That's the simple definition. But this apparently simple skill turns out to be quite complex. It ends up being a rather deep thing about how to work on the basis of limited information and cope with an uncertain world, about knowing yourself and your limitations.

Are most of us bad at this?

Yes. The psychologists Daniel Kahneman and Amos Tversky laid the ground for a lot of what we know about judgement and decision-making. One of their findings is that we are incredibly bad at estimating probabilities. I assumed this was pretty much universal and hard, if not impossible, to overcome. So I was surprised to come across occasional islands of high risk intelligence in odd places.

Where were those pockets of genius?

I found them among horse-race handicappers, bridge players, weather forecasters and expert gamblers. You can only be an expert gambler where there is room for skill, like blackjack, poker or sports betting. It is hard to track them down because they shun publicity, and it was hard to get them to trust me, but eventually they did. I interviewed the blackjack team who inspired the film *21*, as well as other blackjack and poker players. What they have in common is they are very disciplined and hardworking. A distinguishing feature of people with this kind of intelligence is that they've had extensive experience of learning the mistakes of being over-confident in one area, and they apply that lesson generally.

Knowing your limits is key, then?

Yes. It doesn't matter if you've got a high level of knowledge about horses in a race: if you don't have corresponding self-knowledge, it is no good, you won't have high risk intelligence.

How do you quantify risk intelligence?

I set up an online test to measure risk quotient, or RQ. It consists of 50 statements, some true, some false, and you have to estimate the likelihood of a statement being true. The average RQ is not high. There are two ways you can have a low RQ. One is by being over-confident, the other by being under-confident. You do find people making the under-confidence mistake, but there are far fewer of them.

Your book presents a rather worrying finding – that doctors have a very low risk intelligence.

Absolutely. In fact, as they get older, they become more confident but no more accurate, which means that their risk intelligence actually declines. One study I looked at showed that when doctors estimated patients had a 90 per cent chance of having pneumonia, only about 15 per cent had the condition, which is a huge degree of over-confidence. Another way of putting it is that they think they know more than they do. One explanation is that doctors have to make so many different decisions about so many different things that they don't get a chance to build up a good model. Maybe if you have to make life-and-death decisions, you feel you have to exude confidence otherwise you'd be too damned scared to do anything.

What mistakes do we make in assessing risk?

The need for closure is a really interesting one. If you have a great need for closure, it means you don't like being in a state of uncertainty – you want an answer, any answer, even if it is the wrong one. On the other extreme, there is a need to avoid closure, where you are constantly seeking more information, so you get stuck in analysis paralysis.

Can we increase our risk quotient?

Absolutely. One way is by being aware of different cognitive biases. Another is to play a personal prediction game. Bet against yourself and estimate probabilities of anything: whether your partner will get home before 6 o'clock or whether it is going to rain, and keep track of them. Expert gamblers are constantly on the lookout for over-confidence, biases and so on. It is hard work, but it means they know themselves pretty well and they don't have illusions. They know their weaknesses.

7
The greatest problems of maths

In May 2000 the Clay Mathematics Institute in New Hampshire published a list of seven particularly intractable mathematical problems, offering a million-dollar reward for the first correct solution to each. Only one of these Millennium Prizes has so far been solved – the other six are still there for the taking.

3D topology: the Poincaré conjecture

When the brilliant French mathematician Henri Poincaré (1854–1912) was not doing maths, he liked to think about the nature of mathematical creativity. Logic, he felt, was important, but it was not enough. 'It is by logic we prove,' Poincaré wrote, 'it is by intuition that we invent.'

Mathematicians often have intuitions, which they call conjectures or hypotheses, but many are remarkably resistant to logical proof. The Poincaré conjecture, published in 1904, is one of these. It is concerned with topology, the study of shapes, spaces and surfaces, and in particular a form known as a 3-sphere. The surface of an ordinary sphere is two-dimensional, known as a 2-sphere, and consists of points all the same distance from a single point in 3D space, the sphere's centre. A 3-sphere's surface is made up of all points the same distance from a single point in 4D space.

Now imagine you were an ant living on an ordinary 2-sphere's surface. You wouldn't know you were living on a sphere. The topology of your world could be anything in three dimensions – a sphere, a doughnut, even a doughnut with a knot tied in it. But there is a way, in theory at least, that you can tell whether your world is spherical. Trace out a loop around your world's surface and shrink it inwards at a steady rate from each point. If it eventually shrinks to a single point, you live on a sphere.

So much for a 2D surface. We live in a universe of three spatial dimensions and, in our universe, it seems that we can shrink any loop we make down to a point. The Poincaré conjecture says that, as with the 2-sphere, the 3-sphere is the only finite 3D space in which every closed loop can be continuously shrunk to a point. If the universe were finite, the Poincaré conjecture would imply that we are living on the surface of a sphere in four dimensions.

Although difficult to visualize, this conjecture captures our most basic intuitions about 3D space. If it is not true, then our intuitive understanding of spaces and shapes is mistaken. But Poincaré could not prove his conjecture – and nor could anyone else for the best part of a century.

Problem solved

Bizarrely, in the meantime the conjecture was proved in five or more dimensions, and in 1982 for a 4-sphere. But the 3D version representing the universe that we live in remained intractable.

Then, in November 2002, Grigori Perelman of the Steklov Institute of Mathematics in St Petersburg, Russia (see Figure 7.1), released the first of a series of papers online containing calculations that appeared to prove the conjecture. He built on a powerful technique, known as the Ricci flow, developed by Richard Hamilton,

FIGURE 7.1 The Russian mathematician Grigori Perelman, pictured here in 1993, finally solved the notorious Poincaré conjecture in 2002 – but refused all prizes and publicity for it.

a mathematician at Columbia University in New York. This could be used to smooth abstract shapes out into their simplest form by moving the points on the surface, but without 'tearing' it.

In November 2003 Hamilton told a packed meeting at the Clay Mathematics Institute, instigators of the Millennium Prizes, that he thought Perelman's proof correct. It was pored over by mathematicians at a conference the following month and, according to Bruce Kleiner of New York University, was 'essentially confirmed, except possibly for some minor points'. A team of crack mathematicians set about checking the details.

By 2006 there was sufficient confidence that Perelman was awarded a Fields Medal for his proof. He refused it. In July 2010 it was a similar story when the Clay Institute awarded him his $1 million for cracking one of the Millennium Prize problems. On the eve of the Clay celebration, the reclusive mathematician told the Russian news agency Interfax that he thought the organized mathematical community was 'unjust' and that he did not like their decisions. Richard Hamilton, he said, deserved as much credit for the proof as he did.

Regardless of the rights and wrongs, the tools Hamilton and Perelman developed will also enable mathematicians to apply the Ricci flow to other problems in higher-dimensional topology. The now proved conjecture may have implications in general relativity, in which matter and energy warp the fabric of space-time, sometimes producing perplexing 'singularities' that are impossible to smooth out.

Fluid flow: the Navier–Stokes problem

The Navier–Stokes equations are used to describe the behaviour of fluids as they run out of a tap or flow over the wing of a commercial jet, which makes them incredibly important to solve. But their mathematical soundness is in question: for certain problems, it's possible that the equations could malfunction to generate incorrect answers, or give no solutions at all.

Solving the 'Navier–Stokes existence and smoothness problem' means sorting out once and for all what is really going on – and then establishing that the equations are a good fit to reality. Many mathematicians have tried – and failed – to find the answer. The most recent promising proof came from Mukhtarbay Otelbayev of the Eurasian National University in Astana, Kazakhstan, who in 2014 claimed a solution, but later retracted it.

Some physicists are now pinning their hopes on a new approach. Strong coupling is a concept used in physics to describe many situations where a system has many moving parts, making it difficult to model their behaviour exactly – electrons in a superconductor, for example, or something as mundane as the jostling of molecules as water boils in a kettle. New advances in understanding strong coupling problems could help to crack the Navier–Stokes equations, too.

Music of the primes: the Riemann hypothesis

When in the early 1900s the mathematician G. H. Hardy faced a stormy sea passage from Scandinavia to England, he took out an unusual insurance policy. Hardy scribbled a postcard to a friend with the words: 'Have proved the Riemann hypothesis'. God, Hardy reasoned, would not let him die in a shipwreck because he would then be feted for solving the most famous problem in mathematics. Hardy survived the trip.

Almost a century later, the Riemann hypothesis is still unsolved. Its glamour is unequalled because it holds the key to the primes, those mysterious numbers that underpin so much of mathematics (see Chapter 4).

The prime numbers are the atoms of the number system, but annoyingly there is no periodic table for them: where they crop up along the number line is maddeningly unpredictable. In the nineteenth century mathematicians brought a little order to this apparent chaos. Rather as after many coin tosses we expect roughly half heads and half tails, the primes get rarer as you look at larger and larger numbers; and this thinning out is predictable. Below a given number x, the proportion of numbers that are primes is about $1/\ln(x)$, where $\ln(x)$ is the natural logarithm of x. For example, about 4 per cent of numbers smaller than 10 billion are prime.

But that 'about' is very vague. Numbers are products of pure logic, and so surely they ought to behave in a precise, regular way. We would at least like to know how far the prime numbers stray from the distribution.

In 1859 Georg Riemann found a vital clue while studying something called the zeta function. This is a particular way of turning one number into another number, like the function 'multiply by 5'. Riemann fed the zeta function complex numbers, numbers made from a real part and an imaginary part (see Chapter 5).

Complex numbers can be visualized as arrayed on a 2D plane, with real numbers on the horizontal axis and imaginary numbers on the vertical axis. Riemann found that certain complex numbers, when plugged into the zeta function, produce the result zero. The few zeros he could calculate lay on a vertical line in the complex plane, and he guessed that, except for a few well-understood cases, all the infinity of zeros should lie exactly on this line.

The really strange thing was that these Riemann zeros seemed to mirror the pattern of deviations of the prime-number distribution from the $1/\ln(x)$ rule. If the zeros really do lie on the critical line, then the primes stray from that distribution exactly as much as a bunch of coin tosses stray from a 50:50 distribution of heads and tails.

This is a startling conclusion. It suggests that each prime number is picked at random with the probability $1/\ln(x)$, almost as if they were chosen with a weighted coin. So to some extent the primes are tamed. Even if we don't know exactly when a prime number will turn up, we can make statistical predictions about them, just as we can about coin tosses.

But we can only do this if Riemann's guess is right. If the zeros don't line up, the prime numbers are much more unruly. And not just that: hundreds of results in number theory now begin, 'If the Riemann hypothesis is true, then…'. If it is not true, those results must all be re-examined.

The problem is, how do you prove something about an infinity of numbers? Researchers have used supercomputers to calculate the first few billion Riemann zeros above the x-axis, and millions of other zeros higher up, and so far they all lie on the critical line. If just one of them did not, the Riemann hypothesis would be killed. But no amount of computer power can prove the hypothesis true: there are always more zeros to check. And there have been instances before in number theory of plausible conjectures that were supported by seemingly overwhelming numerical evidence, but proved to be false.

Quantum connection

Early in the twentieth century mathematicians made a further daring conjecture: that the Riemann zeros could correspond to the energy levels of a quantum mechanical system. Quantum mechanics deals with the behaviour of tiny particles such as electrons. Crucially, its equations work with complex numbers, but the energy of a physical system is always measured by a real number. So energy levels form an infinite set of numbers lying along the real axis of the complex plane – a straight line rather like Riemann's zeros.

This line is horizontal rather than vertical, but it is a simple bit of maths to rotate the Riemann zero line and put it on top of the real line. If the zeros then match up with the energy levels of a quantum system, the Riemann hypothesis is proved. Various attempts have been made to do just that in a variety of quantum systems – but as yet without success.

If the Riemann hypothesis can be proved, then, using the mathematics of the zeta function, we should be able to predict the results of many quantum experiments, such as scattering of very high energy levels in atoms, molecules and nuclei. It turns out, too, that the same mathematics applies to any situation where waves, including light waves and sound, bounce around chaotically. So the performance of microwave cavities and fibre-optic cables could be improved, and the acoustics of real concert halls might even profit from the music of the primes.

Computational complexity: P = NP?

'Dear Fellow Researchers, I am pleased to announce a proof that P is not equal to NP, which is attached in 10 pt and 12 pt fonts.' So began an email sent in August 2010 to a group of leading computer scientists by Vinay Deolalikar, a mathematician at Hewlett-Packard Labs in Palo Alto, California.

It was an incendiary claim. Deolalikar was saying that he had cracked the biggest problem in computer science, a question concerning the fundamental limits of computation. Talk of the limits of computation tends to be about how many transistors, the building blocks of microprocessors, we can cram on to a silicon chip or whatever material or technology might replace it. The

P = NP? problem raises the spectre that there is a more fundamental limitation, one that lies in the mechanics of computation
itself.

Although Deolalikar's proof was initially promising, an army
of researchers working in an informal online collaboration
soon exposed fundamental flaws in it. 'It's turned out to be an
incredibly hard problem,' says Stephen Cook of the University
of Toronto, Canada, the computer scientist who first formulated P = NP? in May 1971. Today it looks as difficult as ever.

Understanding what the problem is about, and why it is so
important, means breaking it down into its component parts.

What is P?

P and NP are examples of 'complexity classes', categories into
which problems can be slotted depending on how hard they are
to unravel using a computer. P problems are the easy ones: an
algorithm exists to solve them in a 'reasonable' amount of time.
Examples include looking for a number in a list: you check
each number in turn until you find the right one. If the list has
n numbers – the 'size' of the problem – this algorithm takes at
most n steps to search it, so its complexity is proportional to n.
That counts as reasonable.

So, too, does the manual multiplication of two n-digit numbers, which takes about n^2 steps. Any problem of size n whose
solution requires n to the power of something (n^x) steps is relatively quick to crack. It is said to be solvable in 'polynomial
time', and is denoted P.

What is NP?

In some cases, as the size of the problem grows, computing time
increases not polynomially, as n^x, but exponentially, as x^n. This is

a much steeper increase. Imagine, for example, an algorithm to list out all possible ways to arrange the numbers from 1 to n. It is not difficult to envisage the solutions, but the time required to list them rapidly runs away from us as n increases. Even proving that a problem belongs to this non-polynomial class can be difficult, because you have to show that absolutely no polynomial-time algorithm exists to solve it.

With some problems that are difficult to solve in a reasonable time, inspired guesswork might still lead you to an answer whose correctness is easy to verify. Working out a solution to a Sudoku puzzle, for example, can be fiendishly difficult, even for a computer, but presented with a completed puzzle, it is easy to check that it fulfils the criteria for being a valid answer (see Figure 7.2). Problems whose solutions are hard to come by but can be verified in polynomial time make up the complexity class called NP, which stands for non-deterministic polynomial time.

And here is the nub of the P = NP? problem. All problems in the set P are also in the set NP: if you can easily find a solution, you can easily verify it. But is the reverse true? If you can easily verify the solution to a problem, can you also easily solve it – is every problem in the set NP also in P?

What if P ≠ NP?

In 2002 William Gasarch, a computer scientist at the University of Maryland, asked 100 of his peers what they thought the answer to the P = NP? question was. P ≠ NP was the overwhelming winner, with 61 votes. Only nine people supported P = NP – some, they said, just to be contrary. The rest either had no opinion or deemed the problem impossible to solve.

If the majority turn out to be correct, some problems are by their nature so involved that we will never be able to crunch through them. Most computer scientists already assume this to

Constructing a valid Sudoku grid is an example of a computational problem known as a boolean satisfiability problem.

Satisfiability problems are 'NP-hard': as the size of the problem increases, it takes far more computational muscle to find a solution than to check it.

For a 1×1 grid the (only possible) solution is trivial.

For a 4×4 grid, generating a viable solution still takes little computational effort.

A 9×9 grid takes considerably more effort to construct – but it is still relatively easy to check the solution is right.

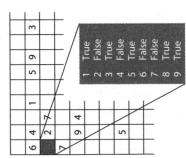

Given an incomplete Sudoku grid, finding a viable solution amounts to evaluating a boolean 'true' or 'false' answer for each empty square – whether each number from 1 to 9 can fit there – and solving it iteratively until no ambiguities remain.

FIGURE 7.2 Constructing a valid Sudoku grid is an example of a computationally 'hard' problem, known as a satisfiability problem.

be the case and concentrate on designing algorithms to find approximate solutions that will suffice for most practical purposes (see Chapter 8). A proof that P ≠ NP would confirm that this is the best we can hope for. It might also shed light on the performance of the latest computing hardware, which splits computations across multiple processors operating in parallel. With twice as many processors, things should run twice as fast, but for certain types of problem they do not. That implies some kind of limitation to computation.

What if P = NP?

If P = NP, the revolution is upon us. That is because of the existence, proved by Cook in his seminal 1971 paper, of a subset of NP problems known as NP-complete. They are the executive key to the NP washroom: find an algorithm to solve an NP-complete problem, and it can be used to solve any NP problem in polynomial time.

Many real-world problems are known to be NP-complete. Satisfiability problems like Sudoku are one example, and so is the notorious travelling salesman problem, which aims to find the shortest-distance route for visiting a series of points and returning to the starting point, an issue of critical interest in logistics and elsewhere (see Figure 7.3).

If we could find a polynomial-time algorithm for any NP-complete problem, it would prove that P = NP, since all NP problems would then be easily solvable. The existence of such a universal computable solution would allow the perfect scheduling of transport, the most efficient distribution of goods and manufacturing with minimal waste. It could also lead to algorithms that perform near-perfect speech recognition and language translation, and that let computers process visual information as well as any human can.

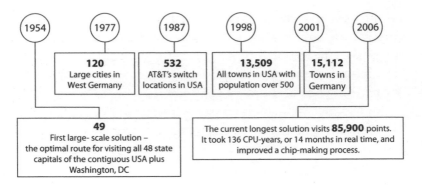

FIGURE 7.3 Solving the travelling salesman problem – how to find the shortest route linking multiple locations – is a computationally hard problem. As computers have improved, so has the size of the problems successfully tackled.

It might also cause online shopping to break. Encryption methods used to protect our personal and bank details in transmission rely on the assumption that breaking a number down to its prime-number factors is hard (see Chapter 4). This certainly looks like a classic NP problem: finding the prime factors of 304,679 is hard, but it is easy enough to verify that they are 547 and 557 by multiplying them together.

A further curious side effect would be that of rendering mathematicians redundant as mathematics became largely mechanizable. Because finding a mathematical proof is difficult, but verifying one is relatively easy, in some way maths itself is an NP problem. If P = NP, we could leave computers to churn out new proofs.

What if there's no algorithm?

There is an odd wrinkle in visions of a P = NP world: that we might prove the statement to be true but never be able to take advantage of that proof. Mathematicians sometimes find 'non-constructive proofs' in which they show that a mathematical object exists without actually finding it. So they could show that an unknown P algorithm exists to solve a problem thought to be NP.

There would be a similar agonizing limbo if a proof that P = NP is achieved with a universal algorithm that scales in complexity as n to the power of a very large number. Being polynomial, this would qualify for the Clay Institute's $1 million prize, but in terms of computability it would not amount to anything.

Particle theory: the Yang–Mills gap

Yang–Mills theory provides a mathematical basis for our existing understanding of elementary particle physics. Without it, we would be unable to say how many particles there are or what masses they should have. But there's a problem. Experiments such as the Large Hadron Collider at CERN near Geneva, Switzerland, as well as computer simulations, suggest that there is a minimum mass that particles can have – you cannot just conjure up a new one with arbitrarily low mass. But the distance between this mass and zero – what is called the 'mass gap' – does not appear to be contained within the framework of Yang–Mills theory. Solving the problem involves mathematically justifying the existence of this gap, with little progress so far to report.

Elliptic curves: the Birch and Swinnerton-Dyer conjecture

Equations known as elliptic curves describe wiggly shapes on a graph, and take the form $y^2 = x^3 + ax + b$, where x and y are variables and a and b are fixed constants. They are used in cryptography and were essential in solving another long-standing and recently solved problem, Fermat's last theorem (see below). Mathematicians who work with these curves use another equation called the L-series to study their behaviour. The Birch and Swinnerton-Dyer conjecture says that if an elliptic curve has an infinite number of solutions, its L-series should equal 0 at certain points. Proving this is true would let mathematicians dive even deeper into these sorts of equations, although the practical applications are not immediately obvious.

Fermat's last theorem

Fermat's last theorem would undoubtedly have been on the Millennium Prize list of the greatest outstanding problems in mathematics, had it not been proved just a few years earlier.

The theorem was a seemingly simple puzzle posed by the seventeenth-century mathematician Pierre de Fermat. He said that for any three whole numbers, a, b and c, the equation $a^n + b^n = c^n$ could not be satisfied by any whole number n greater than 2. We know it is satisfied for $n = 2$, by what are called Pythagorean triples. Set a, b and c equal to 3, 4 and 5, for instance, and the equation works – and you can construct a right-angled triangle whose sides have just those lengths.

Fermat's theorem amounts to saying that similarly satisfying geometric shapes do not exist in more than two dimensions. Tantalizingly, he claimed to have a proof for the theorem that could not be contained in the narrow margin of the textbook he happened to be scribbling in.

His idle musing set mathematicians on a centuries-long quest. It seemingly came to an end in 1993, when British mathematician Andrew Wiles (see Figure 7.4) published a lengthy proof that Fermat was correct. He had worked on the problem in secret for seven years, opening up grand vistas in number theory and developing new tools to tackle other modern mathematical problems that Fermat could not possibly have known about.

That first proof proved to have a few errors but, with the help of colleagues, Wiles was able to announce a new complete version in 1994, which was officially published in the journal *Annals of Mathematics* in 1995.

Wiles was a reluctant celebrity when his proof first hit the headlines, but it is a role he has increasingly accepted. 'In the years since then, I have encountered so many people who told me they have entered mathematics because of the publicity surrounding that and the idea that you could spend your life on these exciting problems, that I've realized how valuable it actually it is,' he says.

In 2016 the Norwegian Academy of Sciences and Letters awarded Wiles the Abel Prize, often called the Nobel Prize of mathematics, 'for his stunning proof of Fermat's last theorem by way of the modularity conjecture for semistable elliptic curves, opening a new era in number theory'.

FIGURE 7.4 Andrew Wiles became the world's most famous
mathematician in 1993, when he published a proof of Fermat's notorious
last theorem.

Higher dimensions: the Hodge conjecture

Mathematicians often find that they can turn a problem in
one field, such as algebra, into one from another field, such as
geometry, to help them solve it. That is what's going on when
you sketch an equation on to a piece of paper as a graph. Such
graphs are two-dimensional, meaning that the related equations
can only have two variables. So how do you use this trick on
equations with three, four or even more variables? The answer
lies in the field of algebraic geometry, which generalizes this
translation idea to higher dimensions.

Algebraic geometers work with techniques and concepts far more complex than simple equations and graphs, and have gradually figured out how to translate problems between them. The Hodge conjecture, the last of the seven Millennium Prize problems, describes how you might do this for a particular type of mathematical object called a Hodge cycle – but until someone proves it right and claims the prize money, we will never know for sure whether you can.

The importance of proof

Proof sets mathematics apart from the other sciences. Mathematics is not an evolutionary subject in which new facts change our interpretations and only the fittest theory survives. Instead, it is like a huge pyramid, with each generation building on the secure foundations of the past.

Ask a mathematician what makes a proof, and they are likely to tell you that it must be absolute – an exhaustive sequence of logical steps leading from an established starting point to an undeniable conclusion. But you cannot just state something you believe is true and move on; you have to convince others that you have made no mistakes. In other words, a proof must be verified.

Verifying a truly ground-breaking proof today can be a frustrating experience. In 2012, for instance, the highly respected mathematician Shinichi Mochizuki of Kyoto University in Japan published more than 500 pages of dense maths on his website. The culmination of years of work, it allowed him to prove a long-standing conundrum about the nature of numbers known as the ABC conjecture. Or so we think: though many mathematicians hailed the result at the time, nobody as yet has been able to verify it.

Recently, mathematicians have also had to come to terms with the possibility of a new kind of error: computer-programming mistakes. Take the four-colour theorem, a suggestion that any map can be shaded using just four colours without any borders sharing the same two colours. You can try this as many times as you like and find it to be true but, to prove it, you need to rule out the very possibility of there being a bizarre map that bucks the trend. In 1976 Kenneth Appel and Wolfgang Haken seemingly did just that. They showed that you could narrow the problem down to 1936 sub-arrangements that might require five colours. They then used a computer to check each of these potential counter-examples, and found that all could indeed be coloured with just four colours – the first major theorem to be proved with the help of a computer.

Only it wasn't, entirely. In 2005 Georges Gonthier of Microsoft Research in Cambridge, UK, and colleagues updated the proof of the four-colour theorem, making every part of it computer-readable. They discovered that a part of the proof, widely assumed to be true because it seemed so obvious, had in fact never been proved at all because it was deemed not worth the effort. Fortunately, the assumption turned out to be correct, but it illustrates the less than absolute nature of some complex modern proofs.

Computers and brute force can get you only so far, too. Consider one of the great unsolved problems about prime numbers: is it possible to generate a sequence of numbers of any length you want in which all the numbers are prime numbers and the difference between all the numbers is the same?

For example, 3, 5, 7 is an arithmetic progression of prime numbers of length three. For a progression of length four, you could take 5, 11, 17, 23, four primes that differ by 6. If you start at the prime 56,211,383,760,397 and count on 44,546,738,095,860 you get another prime. Count on another 44,546,738,095,860 and you get a third prime. If you keep doing this you get 23 primes in an arithmetic progression. This was discovered (using a computer) in 2004 by Markus Frind, Paul Jobling and Paul Underwood.

However, none of this constitutes a proof that you can get any length of sequence you want – and because there are infinitely many prime numbers, this would be impossible by brute force. So there is a still a role for a mathematician's application of logical reasoning. In 2004 Terence Tao of the University of California, Los Angeles, in collaboration with Ben Green of the University of Bristol, laid the issue to rest beautifully, proving that it is theoretically possible to find a suitable sequence of any length you want somewhere in the universe of numbers – a breakthrough that won Tao a Fields Medal in 2006.

Compared with many proofs, Tao and Green's proof is far from arduous in its length and complexity, but they needed about 50 pages and relied on the proofs of many other authors. The only frustrating thing is that it is a non-constructive proof: it tells you that the sequences exist, but not how to find them.

Kepler's stacking problem

The basic problem is familiar to greengrocers everywhere: what is the best way to pile a collection of spherical objects, such as a display of oranges for sale? It shows the lengths mathematicians must sometimes go to, even with a seemingly simple problem, to find an answer that stacks up.

Astronomer and mathematician Johannes Kepler set the ball rolling in 1611, suggesting that a pyramid arrangement was the most efficient. But he could not prove his conjecture. It remained that way until 1998, when Thomas Hales of the University of Pittsburgh, Pennsylvania, presented a proof that Kepler's intuition was correct. Although there are infinite ways to stack infinitely many spheres, most are variations on only a few thousand themes. Hales broke the problem down into the thousands of possible sphere arrangements that mathematically represent the infinite possibilities, and used software to check them all.

That wasn't the end of the story, however. Hales's proof was 300-pages long and took 12 reviewers four years to check for errors. Even when it was published in the journal *Annals of Mathematics* in 2005, the reviewers could say only that they were '99 per cent certain' it was correct.

So, in 2003, Hales started the Flyspeck project, an effort to automatically vindicate his proof through a process known as formal verification. His team used two software assistants called Isabelle and HOL Light, both of them built on a small kernel of logic that has been intensely scrutinized for errors. This provided a foundation that ensured that the computer could check any series of logical statements to confirm they were true. This technology effectively cuts the mathematical referees out of the verification process. As a result, Hales believes their opinions about the correctness of the proof no longer matter.

In 2014 the Flyspeck team announced that they had finally translated the dense mathematics of Hale's proof into computerized form, and verified that it was indeed correct. Hales said at the time that he felt an enormous burden had been lifted from his shoulders. The rest of the mathematical world took a little longer to be convinced – but in 2017 the formal proof was finally accepted in the *Forum of Mathematics* journal.

8
Everyday maths

Mathematics isn't just about high-flown ideas. It can be applied in all sorts of unexpected situations in our everyday lives, to solve problems both profoundly useful and profoundly frivolous.

The algorithm that runs the world

If you have ever wondered how a supermarket supplies the food we have to eat, or how the train or bus that takes you to work is scheduled, the answer may lurk in a server basement somewhere – with an algorithm that is right now working on aspects of your life tomorrow, next week, in a year's time. It is known as the simplex algorithm, and it is used in countless situations where a problem needs to be analysed in many dimensions.

For a mathematician, dimensions are not just about space. To be sure, the concept arose because we have three coordinates of location that can vary independently: up–down, left–right and forwards–backwards. Throw in time, and we have a fourth dimension that works very similarly, except that for reasons unknown we can move through it in only one direction.

Beyond motion, though, we often encounter real-world situations where we can vary many more than four things independently. Suppose, for instance, you are making a sandwich for lunch. Your fridge contains ten ingredients that can be used in varying quantities: cheese, chutney, tuna, tomatoes, eggs, butter, mustard, mayonnaise, lettuce, hummus. These ingredients are the dimensions of a sandwich-making problem, one that can be treated geometrically. Combine your choice of ingredients in any particular way, and your completed snack is represented by a single point in a ten-dimensional space.

In this multi-dimensional space, you are unlikely to have unlimited freedom of movement. There might be only two mouldering hunks of cheese in the fridge, or the merest of scrapings at the bottom of the mayonnaise jar. Your personal preferences might supply other, more subtle constraints to your sandwich-making problem: an eye on the calories, perhaps, or a desire not to mix tuna and hummus. Each of these constraints represents a boundary to our multi-dimensional space beyond which we cannot move.

In business, government and science, similar optimization problems crop up everywhere and quickly morph into brutes with many thousands or even millions of variables and constraints. A fruit importer might have a 1000-dimensional problem to deal with, for instance, shipping bananas from five distribution centres storing varying numbers of fruit to 200 shops each with a need for different numbers. How many items of fruit should be sent from which centres to which shops while minimizing total transport costs?

A fund manager might, similarly, want to arrange a portfolio optimally to balance risk and expected return over a range of stocks; a railway timetabler to decide how best to roster staff and trains; or a factory or hospital manager to work out how to juggle finite machine resources or ward space. Each such problem can be depicted as a geometrical shape, a 'polytope', whose number of dimensions is the number of variables in the problem, and whose boundaries are delineated by whatever constraints there are.

The simplex algorithm provides a way to box our way through a polytope towards its optimal point. It emerged in the late 1940s from the work of the US mathematician George Dantzig, who had spent the Second World War investigating ways to increase the logistical efficiency of the US Air Force.

One of the first insights he arrived at was that the optimum value of the 'target function' – the thing we want to maximize or minimize, be that profit, travelling time or whatever – is guaranteed to lie at one of the corners of the polytope (see Figure 8.1). This instantly makes things much more tractable: there are infinitely many points within any polytope, but only ever a finite number of corners.

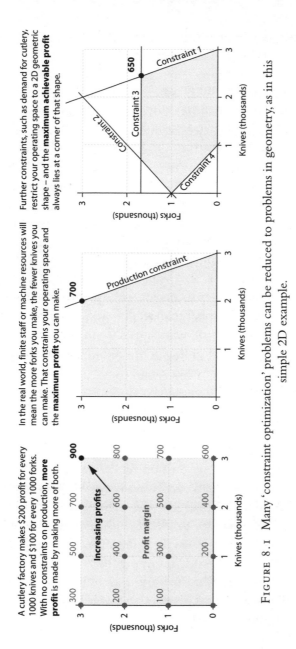

A cutlery factory makes $200 profit for every 1000 knives and $100 for every 1000 forks. With no constraints on production, **more profit** is made by making more of both.

In the real world, finite staff or machine resources will mean the more forks you make, the fewer knives you can make. That constrains your operating space and the **maximum profit** you can make.

Further constraints, such as demand for cutlery, restrict your operating space to a 2D geometric shape – and the **maximum achievable profit** always lies at a corner of that shape.

FIGURE 8.1 Many 'constraint optimization' problems can be reduced to problems in geometry, as in this simple 2D example.

The bad news, however, is that the number of corners – though finite – may be astronomically large. Even just a ten-dimensional problem with 50 constraints – perhaps trying to assign a schedule of work to ten people with different expertise and time constraints – may already land us with several billion corners to try out. Rather than testing each of them in turn, the simplex algorithm finds a quicker way through. It implements a 'pivot rule' at each corner. Subtly different variations of this pivot rule exist in different implementations of the algorithm, but often it involves picking the edge along which the target function descends most steeply, ensuring that each step takes us nearer the optimal value. When a corner is found where no further descent is possible, we know we have arrived at the optimal point.

The simplex method is generally a very slick problem-solver indeed, typically reaching an optimum solution after a number of pivots comparable to the number of dimensions in the problem. That means a likely maximum of a few hundred steps to solve a 50-dimensional problem, rather than billions with a brute-force approach. Such a running time is said to be 'polynomial' or simply 'P', the benchmark for practical algorithms that have to run on finite processors in the real world (see Chapter 7).

Dantzig's algorithm saw its first commercial application in 1952, to discover how best to blend available stocks of four different petroleum products into an aviation fuel with an optimal octane level. Since then, the simplex algorithm has steadily conquered the world, embedded both in commercial optimization packages and bespoke software products. Probably tens or hundreds of thousands of calls of the simplex method are made every minute.

For mathematicians, it is still not a perfect solution, however. To start with, the algorithm's running time is polynomial only on average. However well it does in general, it always seems possible to concoct some awkward optimization problems in which it performs poorly. The good news is that these path-ological cases tend not to show up in practical applications, although exactly why this should be so remains unclear.

A pretender came on the scene in the 1970s and 1980s with the discovery of 'interior point methods', flashy algorithms that, rather than feeling their way around a polytope's surface, drill a path through its core. They came with a genuine mathematical seal of approval – a guarantee always to run in polynomial time – and typically took fewer steps to reach the optimum point than the simplex method, rarely needing more than 100 moves regardless of how many dimensions the problem had.

The trouble with interior point methods is that each step entails far more computation than a simplex pivot: instead of comparing a target function along a small number of edges, you must analyse all the possible directions within the poly-tope's interior, a gigantic undertaking. For some huge indus-trial problems, this trade-off is worth it, but for by no means all. Jacek Gondzio, an optimization specialist at the University of Edinburgh, UK, estimates that between 80 and 90 per cent of today's linear optimization problems are still solved by some variant of the simplex algorithm, as well as a good few of the less common, but even more complex, non-linear problems. Gondzio admits to being a devoted interior-point researcher. 'I'm doing my best trying to compete,' he says.

We would still dearly love to find something better: some new variant of the simplex algorithm that preserves all its advantages but also invariably runs in polynomial time. Yet even the existence of such an algorithm depends on a fundamental geometrical assumption – that a short enough path around the surface of a polytope between two corners actually exists. This conjecture was first formulated by the US mathematician and pioneer of operations research Warren Hirsch in 1957, in

2000 years of algorithms

George Dantzig's simplex algorithm has a claim to be the world's most significant. But algorithms go back much further.

c.300 BCE The Euclidean Algorithm

From Euclid's mathematical primer *Elements,* this is the very first of all algorithms, showing how, given two numbers, you can find the largest number that divides into both. It has still not been bettered.

1946 The Monte Carlo Method

When your problem is just too hard to solve directly, enter the casino of chance. John von Neumann, Stanislaw Ulam and Nicholas Metropolis's Monte Carlo algorithm taught us how to play – and win.

1957 The Fortran Compiler

Programming was a fiddly, laborious job until an IBM team led by John Backus invented the first high-level programming language, Fortran. At the centre is the compiler: the algorithm that converts the programmer's instructions into machine code.

1994 Shor's Algorithm

Bell Labs's Peter Shor found a new, fast algorithm for splitting a whole number into its constituent primes – but it could only be performed by a quantum computer. If ever implemented on a large scale, it would nullify almost all modern Internet security.

1998 Pagerank

The Internet's vast repository of information would be of little use without a way to search it. Stanford University's Sergey Brin and Larry Page found a way to assign a rank to every web page – and the founders of Google have been living off it ever since.

820 CE The Quadratic Algorithm

The word 'algorithm' is derived from the name of the Persian mathematician al-Khwārizmī. Experienced practitioners today perform his algorithm for solving quadratic equations (those containing an x^2 term) in their heads. For everyone else, modern algebra provides the formula familiar from school.

1936 The Universal Turing Machine

British mathematician Alan Turing equated algorithms with mechanical processes – and found one to mimic all the others, the theoretical template for the programmable computer.

1962 Quicksort

Extracting a word from the right place in a dictionary is an easy task; putting all the words in the right order in the first place is not. The British mathematician Tony Hoare provided the recipe, now an essential tool in managing databases of all kinds.

1965 The Fast Fourier Transform

Much digital technology depends on breaking down irregular signals into their pure sine-wave components – making James Cooley and John Tukey's algorithm one of the world's most widely used.

a letter to Dantzig musing on the efficiency of the simplex algorithm.

The Hirsch conjecture was proved true for all 3D polyhedra in 1966, but there has always been a hunch that the same did not hold for higher-dimensional polytopes. And in 2010 the Hirsch conjecture was proved false by mathematician Francisco Santos, albeit for a single case of a 43-dimensional polytope with 86 faces. According to Hirsch's conjecture, the longest path across this shape would have (86 − 43) steps, that is, 43 steps. Santos was able to establish conclusively that it contains a pair of corners at least 44 steps apart.

Since Santos's first disproof, further Hirsch-defying polytopes have been found in dimensions as low as 20. The only known limit on the shortest distance between two points on a polytope's surface is much larger than the one the Hirsch conjecture would have provided − far too big, in fact, to guarantee a reasonable running time for the simplex method, whatever fancy new pivot rule we might dream up. If this is the best we can do, the goal of an idealized algorithm will remain forever out of reach, with potentially serious consequences for the future of optimization.

A highly efficient variant of the simplex algorithm may still be possible if the so-called polynomial Hirsch conjecture is true. This sets a weaker standard, guaranteeing only that no polytopes have paths disproportionately long compared with their dimension and number of faces, but there is as yet no conclusive sign that it is true, either.

How to slice a pizza

Be it cake, pizza or anything else, we have probably caught ourselves eyeing the biggest slice − or quietly bemoaning the fact that life has served us one too small. In the early 1990s the question of fair pizza slicing became an obsession for mathematicians Rick Mabry and Paul Deiermann, then both at Louisiana State University in the USA. Mabry recalls that the pair went to lunch together at least once a week and one of them would always bring a notebook to draw pictures, while their food got cold.

Suppose, for example, a harried waiter cuts a pizza off-centre, but with all the edge-to-edge cuts crossing at a single point, and with the same angle between adjacent cuts. The off-centre cuts mean that the slices will not be all the same size. If two people take turns to take neighbouring slices, will they get equal shares by the time they have gone right round the pizza. If not, who will get more?

You could estimate the area of each slice, tot them all up and then work out each person's total from that. But a mathematician's aim is to distil any problem down to a few general, provable rules that avoid exact calculations and that work every time. The easiest example is when at least one cut passes plumb through the centre of the pizza. The pieces then pair up on either side of the cut through the centre, and so can be divided evenly between two diners, no matter how many cuts there are.

But what if none of the cuts passes through the centre? For a pizza cut once, the answer is obvious by inspection: whoever eats the centre eats more. It is the same answer for a pizza cut twice, yielding four slices. But that turns out to be an anomaly. The three general rules that deal with greater numbers of cuts emerged only over subsequent years to form the complete pizza theorems.

The first rule proposes that if you cut a pizza through the chosen point with an even number of cuts more than 2, the pizza will be divided evenly between two diners who each take alternate slices. With an odd number of cuts, things start to get more complicated. Here, the pizza theorem says that if you cut the pizza with 3, 7, 11, 15 ... cuts, and no cut goes through the centre, then the person who gets the slice that includes the centre of the pizza eats more in total. If you use 5, 9, 13, 17 ... cuts, the person who gets the centre ends up with less (see Figure 8.2).

Rigorously proving this to be true, however, proved a tough challenge. Deiermann quickly sketched a solution to the three-cut problem. The pair went on to prove the statement for five

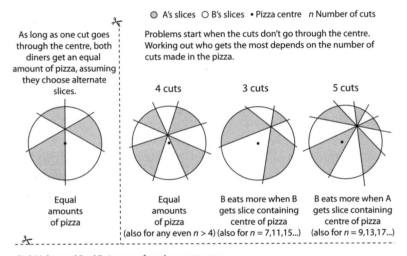

○ A's slices ○ B's slices • Pizza centre *n* Number of cuts

As long as one cut goes through the centre, both diners get an equal amount of pizza, assuming they choose alternate slices.

Problems start when the cuts don't go through the centre. Working out who gets the most depends on the number of cuts made in the pizza.

	4 cuts	3 cuts	5 cuts

Equal amounts of pizza

Equal amounts of pizza
(also for any even *n* > 4)

B eats more when B gets slice containing centre of pizza
(also for *n* = 7,11,15...)

B eats more when A gets slice containing centre of pizza
(also for *n* = 9,13,17...)

Rick Mabry and Paul Deiermann found a way to prove the pizza conjecture that involves comparing opposite slices in turn.

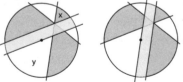

Instead of looking at the actual slices (*x* and *y*, say), they drew a line parallel to each cut running through the centre of the pizza

They then used the 'rectangular' light-grey areas as a measure of the difference in area of opposing slices. Plug that in to some complicated algebra and the proof arises.

FIGURE 8.2 The 'pizza conjecture' asks who will get the biggest portion of pizza, assuming that diners A and N take alternate slices and that the angles between adjacent cuts are all equal.

cuts, and then proved that if you cut the pizza seven times, you get the same result as for three cuts: the person who eats the centre of the pizza ends up with more.

Boosted by their success, they thought they might have stumbled across a technique that could prove the entire pizza theorem once and for all. For an odd number of cuts, opposing slices inevitably go to different diners, so an intuitive solution is to simply compare the sizes of opposing slices and figure out who gets more, and by how much, before moving on to the next pair. Working your way around the pizza pan, you tot up the differences and there's your answer.

In practice, though, it is extremely difficult to come up with a solution that covers all the possible numbers of odd cuts. The two hoped they might be able to deploy a deft geometrical trick to simplify the problem. The key was the area of the rectangular strips lying between each cut and a parallel line passing through the centre of the pizza. Mabry says the formula for the area of strips is easier than for slices, and the strips have the added benefit of giving some very nice visual proofs of certain aspects of the problem. Unfortunately, the solution still included a complicated set of sums of algebraic series involving tricky powers of trigonometric functions. These took 11 years to figure out.

The breakthrough came in 2006, when Mabry was on a vacation in Kempten im Allgäu in the far south of Germany without a computer. Putting technology aside, he managed to refashion the algebra into a manageable, more elegant form. Back home, he suspected that someone, somewhere must already have worked out the simple-looking sums at the heart of the new expression, so he trawled the online world for theorems in the vast field of combinatorics – an area of pure mathematics concerned with listing, counting and rearranging – that might provide the key result he was looking for. Eventually, he found what he was after: a 1999 paper that referenced a mathematical statement from 1979. The rest of the proof then fell into place, and it was finally published in 2009.

Besides fair slicing of pizza and other circular treats, the pizza theorem has no obvious applications yet – but that's not the point. 'It's a funny thing about some mathematicians,' says Mabry. 'We often don't care if the results have applications because the results are themselves so pretty.' And to be fair, sometimes solutions to abstract mathematical problems do show their face in unexpected places. For example, a nineteenth-century mathematical curiosity called the 'space-filling curve' recently resurfaced as a model for the shape of the human genome.

Other mathematicians have since taken up the pizza cutter, coming up, for example, with a recipe for slicing that gives you 12 identically shaped pieces, six of which form a star extending out from the centre, while the other six divide up the crusty remainder. This is perfect if the centre of the pizza has a topping that some people would rather avoid, while others desperately want crust for dipping (see Figure 8.3).

In 2015 Joel Haddley and Stephen Worsley of the University of Liverpool, UK, generalized the technique to create even more ways to slice. They proved that you can create similar tilings from curved pieces with any odd number of sides – known as 5-gons, 7-gons and so on – dividing them in two as before. Mathematically, there is no limit whatsoever, says Haddley, although you might find it impractical to carry out the scheme beyond 9-gon pieces.

Like Mabry, Haddley is sanguine about applications of his work outside pizza cutting. For him, it's interesting mathematically, and it produces some tasty pictures.

FIGURE 8.3 Pizza-slicing algorithms can produce some complex patterns.

Beating the odds

When Edward Thorp, a mathematics student at the Massachusetts Institute of Technology, stood at the roulette wheel in a Las Vegas casino in the summer of 1961, he knew pretty well where the ball was going to land. He walked away with a profit, took it to the racecourse, the basketball court and the stock market, and became a multimillionaire. He was not on a lucky streak: he had harnessed his knowledge of mathematics to understand, and beat, the odds.

Thorp was armed with the first 'wearable' computer, one that could predict the outcome of the spin. Once the ball was in play, Thorp fed the computer information about the speed and position of the ball and the wheel using a microswitch inside his shoe. It made a forecast about a probable result, and Thorp bet on neighbouring numbers. Thorp's device would now be illegal in a casino. But knowledge of the workings of probability can help anyone beat the odds in various games – well, sort of.

Roulette

There is a simple and sure-fire way to win at the roulette table – as long as you have deep pockets. A spin of the roulette wheel is just like the toss of a coin. Each spin is independent, with a 50:50 chance of the ball landing on black or red. Contrary to intuition, a black number is just as likely to appear after a run of 20 consecutive black numbers as the seemingly more likely red. This false intuition is known as the gambler's fallacy.

So always bet on the same colour. If you lose, double your bet on the next spin. Because your colour will come up eventually, this method will always produce a profit. The downside is that you'll need a big pot of cash to stay in the game. A losing streak can escalate your bets very quickly: seven unlucky spins on a £10 starting bet will have you parting with a hefty £1,280 on the eighth. Unfortunately, your winnings don't escalate in the same way: when you do win, you will only make a profit equal to your original stake. So while the theory itself is sound, the roulette wheel is likely to keep on taking your money longer than you can remain solvent.

Blackjack

In a game such as blackjack you can tip the odds in your favour simply by keeping track of the cards – within the rules, if not the spirit, of a game of chance.

Blackjack starts with each player being dealt two cards face up. Face cards count as 10 and the ace as 1 or 11 at the player's discretion. The aim is to have as high a total as possible without busting – going over 21. To win, you must achieve a score higher than the dealer's. Cards are dealt from a shoe, a box of cards made up of three to six decks. Players can stick with the two cards they are dealt or 'hit' and receive an extra card to try to get closer to 21. If the dealer's total is 16 or less, the dealer must hit. At the end of each round, used cards are discarded.

The basic idea of card counting is to keep track of those discarded cards to know what's left in the shoe. A shoe rich in high cards will slightly favour you, while a shoe rich in low cards is slightly better for the dealer. With lots of high cards still to be dealt, you are more likely to score 20 or 21 with your first two cards, and the dealer is more likely to bust if his initial cards are less than 17. An abundance of low cards benefits the dealer for similar reasons.

If you keep track of which cards have been dealt, you can gauge when the game is swinging in your favour. The simplest way is to start at zero and add or subtract according to the dealt cards. Add 1 when low cards (two to six) appear, subtract 1 when high cards (10 or above) appear, and stay put on seven, eight and nine. Then place your bets accordingly – bet small when your running total is low, and when your total is high, bet big. This method can earn you a positive return of up to 5 per cent on your investment. It is a lot of effort for a small return – but, in this era of low interest rates, possibly worth a punt.

Lotteries

The evening of 14 January 1995 was one that Alex White will never forget. It was the ninth ever draw of the UK National Lottery, with an estimated jackpot of a massive £16 million, and White (not his real name) matched all six numbers: 7, 17,

23, 32, 38 and 42. Unfortunately, White won only £122,510 because 132 other people chose the same combination of numbers and took a share of the jackpot.

Dozens of methods claim to improve your odds of winning the lottery. None of them works. Every combination of six numbers has the same odds of winning as any other – 1 in 13,983,816 in the 49-ball game White was playing. (Since 2015 players must choose six numbers from 59, giving them a much lower chance of winning the top prize: 1 in 45,057,474). But, as White's story shows, the fact that you could have to share the jackpot suggests a way to maximize any winnings: win with numbers nobody else has chosen.

Shortly after the start of the UK National Lottery in 1994, mathematician Simon Cox of the University of Southampton worked out lottery players' favourite figures by analysing data from 113 lottery draws, comparing the winning numbers with how many people had matched four, five or six of them. Seven was the favourite, chosen 25 per cent more often than the least popular number, 46. Numbers 14 and 18 were also popular, while 44 and 45 were among the least favourite. There was a noticeable preference for numbers up to 31. This is called 'the birthday effect', says Cox, because many people use their date of birth.

Several other patterns emerged. The most popular numbers are clustered around the centre of the form people fill in to make their selection. Similarly, many players appear to just draw a diagonal line through a group of numbers on the form. There is also a clear dislike of consecutive numbers. People refrain from choosing numbers next to each other, even though getting 1, 2, 3, 4, 5, 6 is as likely as any other combination. Numerous studies on the US, Swiss and Canadian lotteries have produced similar findings. Perhaps the most notable feature of White's

popular number choice is that they are relatively evenly distributed – they 'look' random.

To test the idea that picking unpopular numbers can maximize your winnings, Cox simulated a virtual syndicate that
bought 75,000 tickets each week, choosing its numbers at random. Using the real results of the first 224 UK lottery draws,
he calculated that his syndicate would have won a total of £7.5
million – on an outlay of £16.8 million. If his syndicate had
stuck to unpopular numbers, however, it would have more than
doubled its winnings.

It is therefore better to go for numbers above 31, and pick
ones that are clumped together or situated around the edges of
the form. Then, if you match all six numbers, you are less likely
to have to share with others. But bear in mind that probability
also predicts that you are unlikely to win the top prize for many
centuries yet.

Racing

Although it would be nearly impossible to beat a seasoned
bookie at his or her own game, play two or three bookies
against each other and you can come up a winner, whatever
the outcome of a race.

Let's say, for example, you want to bet on one of the highlights of the British sporting calendar, the annual university boat
race between old rivals Oxford and Cambridge. One bookie is
offering 3 to 1 on Cambridge to win and 1 to 4 on Oxford. But
a second bookie disagrees and has Cambridge evens (1 to 1)
and Oxford at 1 to 2.

Each bookie has looked after his or her own back, ensuring
that it is impossible for you to bet on both Oxford and Cambridge with him/her and make a profit regardless of the result.

However, if you spread your bets between the two bookies, it is possible to guarantee success. Having done the calculations, you place £37.50 on Cambridge with bookie 1 and £100 on Oxford with bookie 2. Whatever the result, you make a profit of £12.50.

Guaranteeing a win this way is known as 'arbitrage', but opportunities to do it are rare and fleeting. The fewer runners there are in a race, the better it works. It's not necessarily risk-free because you might not be able to get the bet you want exactly when you need it – but it is enough for some professional gamblers to make a living out of it.

Knowing when to stop

Gambling can be addictive, especially when you get tantalizingly close to finding a winning combination or strategy. And this is a problem even when you have mathematics on your side: it's all too easy to lose sight of what you could lose. Fortunately, that is something probability can help you with, too.

If you have trouble knowing when to quit, try getting your head around 'diminishing returns' – the optimal stopping tool. One way to demonstrate diminishing returns is a thing called the marriage problem. Suppose you are told that you must marry, and that you must choose your spouse out of 100 applicants. You may interview each applicant once. After each interview, you must decide whether to marry that person. If you decline, you lose the opportunity for ever. If you work your way through 99 applicants without choosing one, you must marry the 100th. You may think you have 1 in 100 chance of marrying your ideal partner, but the truth is that you can do a lot better than that.

Interview half the potential partners, then stop at the next best one – that is, the first one better than the best person you've already interviewed. A quarter of the time, the second-best partner will be in the first 50 people and the very best in the second. So 25 per cent of the time, the rule 'stop at the next best one' will see you marrying the best candidate.

You can do even better. John Gilbert and Frederick Mosteller of Harvard University proved that you could raise your odds to 37 per cent by interviewing 37 people then stopping at the next best. The number 37 comes from dividing 100 by e, the base of natural logarithms, which is roughly equal to 2.72 (see Chapter 5). Gilbert and Mosteller's law works no matter how many candidates there are – you simply divide the number of options by e. Suppose you find 50 companies that offer car insurance but you have no idea whether the next quote will be better or worse than the previous one. Should you get a quote from all 50? No, phone up 18 (50 divided by 2.72) and go with the next quote that beats the first 18.

This can also help you decide the optimal time to stop gambling. Decide on the maximum number of bets you will make – 20, for example. To maximize your chance of walking away at the right time, make seven bets and stop at the next one that wins you more than the previous biggest win.

Spaghetti functions

When looking for architectural inspiration, the last place you might think to look is the bottom of a pasta bowl. Architect and designer George Legendre did just that, however, compiling in 2011 the first comprehensive mathematical taxonomy of the stuff.

Pasta has spawned a multiplicity of complex forms around the world – think spaghetti, ravioli, the tube-like penne or butterfly-shaped farfalle. But that obscures an unexpected mathematical simplicity: if you look carefully, there are probably only three basic topological shapes in pasta – cylinders, spheres and ribbons, says Legendre.

It was a late-night glass of wine too many at his architectural practice in London that inspired Legendre, together with his colleague Jean-Aimé Shu, to use mathematics to bring order to this chaotic world. They started by ordering lots of pasta. Then, using their design know-how, they set about modelling every shape they could lay their hands on to derive formulae that encapsulate their forms. This exercise took nearly a year.

For each shape, they needed three expressions, each describing its form in one of the three dimensions. This provides a set of coordinates that, plotted on a graph, faithfully represents the pasta in 3D. The curvaceous shapes of most pasta lend themselves to mathematical representations mainly through oscillating sine and cosine functions. For some pastas, the right recipe was obvious. Spaghetti, for example, is little more than an extruded circle. The sine and cosine of a single angle serves to define the coordinates of the points enclosing its unvarying cross section, and a simple constant characterizes its length. Similarly, grain-like puntalette are just deformed spheres. The sines and cosines of two angles, together with different multiplying factors to stretch the shape out in three dimensions, supply its mathematical likeness.

Other shapes are harder to crack. Scrunched-up saccottini, for example, require a complex mathematical mould of multiplied sines and cosines. Simple features such as the slanted ends of penne take some low modelling cunning, involving chopping the pasta into pieces, each represented by slightly different equations.

Sharp inflections, such as the undulating crests of the cocks-comb-like galletti, are tricky, too, although trigonometric functions again turn out to be the best tools for the job: raising sines and cosines to a higher power constricts the smooth, oscillating shape of the function into something approaching a spike. A similar technique can be used to broaden out the function into something approaching a right angle.

In the end, Legendre had a compendium of 92 pasta shapes, each exactly modelled and divided into categories according to the mathematical relationships revealed between them – some obvious, some less so. The twisted ribbons of sagne incannulate and the 'little hats', cappelletti, turn out to be topologically identical: given sufficiently pliant dough, deft hands could stretch, twist and remould one shape into the other without the intervention of a knife or pair of scissors.

Legendre's pasta taxonomy provides playful proof that immense variety and seeming complexity can be reduced to simple mathematical beginnings. A similar approach might lead to a new, more efficient way of translating design into engineering that is useful for much larger structures. Plans for an arbitrarily complex skyscraper, for example, might be reduced to equations for each of its three dimensions just like those that define the pasta shapes. He sees the equations for a cross section as indicative of a floor, with a third equation for the elevation.

Legendre's Henderson Waves bridge in Singapore (see Figure 8.4) has an undulating form more than a little reminiscent of graceful pasta-like curves, and it was modelled using exactly the same principles. 'I just gave the engineers equations,' he says.

FIGURE 8.4 The curves of the Henderson Bridge in Singapore were influenced by studies of the topology of pasta shapes.

Why democracy is always unfair

In an ideal world, elections should be two things: free and fair. Every adult, with a few sensible exceptions, should be able to vote for a candidate of their choice, and each single vote should be worth the same.

Ensuring a free vote is a matter for the law. Making elections fair is more a matter for mathematicians. The many democratic electoral systems in use around the world attempt to strike a balance between mathematical fairness and political considerations such as accountability and the need for strong, stable government. Mathematicians and others have been studying voting systems for hundreds of years, looking for sources of bias that distort the value of individual votes, and ways to avoid them.

In 1963 the American economist Kenneth Arrow listed the general attributes of an idealized fair voting system. He suggested that voters should be able to express a complete set of their preferences; no single voter should be allowed to dictate the outcome of the election; if every voter prefers one candidate to another, the final ranking should reflect that; and if a voter prefers one candidate to a second, introducing a third candidate should not reverse that preference.

This all sounds very sensible, but there is just one problem: Arrow and others went on to prove that no conceivable voting system could satisfy all four conditions. In particular, there will always be the possibility that one voter, simply by changing their vote, can change the overall preference of the whole electorate. So we are left to make the best of a bad job – and deal with the mathematical imperfections that various electoral systems present.

First past the post

First past the post, or 'plurality', voting is used for national elections in Canada, India, the UK and the USA. Its principle is simple: each electoral division elects one representative, the candidate who gained the most votes.

This system scores well on stability and accountability, but in terms of mathematical fairness it is a dud. Votes for anyone other than the winning candidate are disregarded. If more than two parties with substantial support contest a constituency, as is typical in Canada, India and the UK, a candidate does not have to get anything like 50 per cent of the votes to win, so a majority of votes are 'lost'.

This means that a party can win outright by being only marginally ahead of its competitors in most electoral divisions.

In the UK general election of 2005, for example, the ruling Labour Party won 55 per cent of the seats on just 35 per cent of the total votes. If a candidate or party is slightly ahead in a bare majority of electoral divisions but a long way behind in others, they can win even if a competitor gets more votes overall – as happened most notoriously in recent history with Donald Trump's defeat of Hillary Clinton in the US presidential election of 2016.

In first-past-the-post systems, borders matter. To ensure that each vote has roughly the same weight, each constituency should have roughly the same number of voters. Threading boundaries between and through centres of population on the pretext of ensuring fairness is also a great way to cheat for your own benefit. The practice is known as gerrymandering, after a nineteenth-century governor of Massachusetts, Elbridge Gerry, who created an electoral division whose shape reminded a local newspaper editor of a salamander (see Figure 8.5).

Suppose a city controlled by the Liberal Republican (LR) party has a voting population of 900,000 divided into three constituencies. Polls show that at the next election LR is heading for defeat – 400,000 people intend to vote for it but the 500,000 others will opt for the Democratic Conservative (DC) party. If the boundaries were to keep the proportions the same, each constituency would contain roughly 130,000 LR voters and 170,000 DC voters, and DC would take all three seats – the usual inequity of a plurality voting system.

In reality, voters inclined to vote for one party or the other will probably clump together in the same neighbourhoods of the city, so LR might well retain one seat. However, it could be all too easy for LR to redraw the boundaries to reverse the result and secure itself a majority – as the two dividing strategies in Figure 8.6 show.

FIGURE 8.5 The term 'gerrymandering' was inspired by a convoluted division of voting districts in nineteenth-century Massachusetts.

The anomalies of a plurality voting system can be subtler, though, as demonstrated by mathematician Donald Saari at the University of California, Irvine. Suppose 15 people are asked to rank their liking for milk (M), beer (B) and wine (W). Six rank them M–W–B, five B–W–M and four W–B–M. In a plurality system, where only first preferences count, the outcome is simple: milk wins with 40 per cent of the vote, followed by beer, with wine trailing in last.

Each square represents 100,000 voters

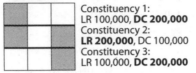

■ Liberal republicans (**LR**): Total votes 400,000

☐ Democratic Conservatives (**DC**): Total votes 500,000

Scenario 1

Constituency 1:
LR 100,000, **DC 200,000**
Constituency 2:
LR 200,000, DC 100,000
Constituency 3:
LR 100,000, **DC 200,000**

LR 1 Seat, DC 2 Seats – **DC wins**

Scenario 2

Constituency 1 (left):
LR 200,000, DC 100,000
Constituency 2 (top):
LR no votes, **DC 300,000**
Constituency 3 (bottom):
LR 200,000, DC 100,000

LR 2 Seats, DC 1 Seat – **LR wins**

FIGURE 8.6 In a first-past-the-post voting system, where boundaries fall can determine the outcome.

So do voters actually prefer milk? Not a bit of it. Nine voters prefer beer to milk, and nine prefer wine to milk – clear majorities in both cases. Meanwhile, ten people prefer wine to beer. By pairing off all these preferences, we see the truly preferred order to be W-B-M – the exact reverse of what the voting system produced. In fact, Saari showed that, given a set of voter preferences, you can design a system that produces any result you desire.

In the example above, simple plurality voting produced an anomalous outcome because the alcohol drinkers stuck together: wine and beer drinkers both nominated the other as their second preference and gave milk a big thumbs-down. Similar things happen in politics when two parties appeal to the same kind of voters, splitting their votes between them and allowing a third party unpopular with the majority to win the election.

Unfortunately, it is only possible to a certain extent to avoid that kind of unfairness while keeping the advantages of a first-past-the-post system. One possibility is a second 'run-off' election between the two top-ranked candidates, as happens in France and in many presidential elections elsewhere. But there is no guarantee that the two candidates with the widest potential support even make the run-off. In the 2017 French presidential election, for example, a splitting of the vote in the traditional left and right blocs meant that the final election was contested between a candidate of the far right, Marine Le Pen, and the liberal insurgent Emmanuel Macron. In 2002 the run-off ended up being contested by a centre-right and a far-right candidate.

Alternative voting

Alternative voting (AV) is a strategy that allows voters to place candidates in order of preference, with a 1, 2, 3 and so on. After the first-preference votes have been counted, the candidate with the lowest score is eliminated and the votes reapportioned to the next-choice candidates on those ballot papers. This process goes on until one candidate has the support of over 50 per cent of the voters. This system, called the instant run-off or alternative or preferential vote, is used in elections to the Australian House of Representatives, as well as in several US cities. It was rejected in a referendum to change the system for parliamentary elections in the UK in 2011.

Preferential voting comes closer to being fair than plurality voting, but it does not eliminate ordering paradoxes. The Marquis de Condorcet, a French mathematician, noted this as early as 1785. Suppose we have three candidates, A, B and C, and three voters who rank them A-B-C, B-C-A and C-A-B. Voters prefer A to B by 2 to 1. But B is preferred to C and C is preferred to A by the same margin of 2 to 1. To quote the Dodo in *Alice's Adventures in Wonderland*: 'Everybody has won and all must have prizes.'

Proportional representation

One type of voting system avoids ordering paradoxes entirely: proportional representation (PR). Here, a party is awarded a number of parliamentary seats in direct proportion to the number of people who voted for it. Such a system is undoubtedly fairer in a mathematical sense than either plurality or preferential voting, but it has political drawbacks. It implies large, multi-representative constituencies; the best shot at truly proportional representation comes with just one constituency, the system used in Israel. But large constituencies weaken the link between voters and their representatives. Candidates are often chosen from a centrally determined list, so voters have little or no control over who represents them.

And proportional representation has its own mathematical wrinkles. There is no way, for example, to allocate a whole number of seats in exact proportion to a larger population. This can lead to an odd situation in which increasing the total number of seats available reduces the representation of an individual constituency, even if its population stays the same.

Elections to the US House of Representatives use a first-past-the-post voting system, but the US Constitution requires that seats be 'apportioned among the several states according to their respective numbers' – that is, divided up proportionally. In 1880 the chief clerk of the US Census Bureau, Charles Seaton, discovered that Alabama would get eight seats in a 299-seat House but only seven in a 300-seat House. This 'Alabama paradox' was caused by an algorithm known as the largest remainder method, which was used to round the number of seats a state would receive under strict proportionality to a whole number.

Suppose, for simplicity's sake, that a nation of 39 million voters has a parliament with four seats – giving a quota of 9.75 million voters per seat. The seats must, however, be shared among three states, Alabaska, Bolorado and Carofornia, with voting populations of 21, 13 and 5 million respectively. Dividing these numbers by the quota gives each state's fair proportion of seats: 2.15, 1.33 and 0.51. Rounded down to an integer, this number of seats is given to the states. Any seats left over go to the state or states with the highest remainders: in this case Carofornia. So Alabaska gets two seats and Bolorado and Carofornia one each.

Suppose now that the number of seats increases from four to five. The quota is 39 million divided by 5, or 7.8 million. The fair proportions for the three states are now 2.69, 1.67 and 0.64. The rounded-down integers account for three seats as before. The two spare go to Alabaska and Bolorado, which have the two largest remainders, and Carofornia loses its only seat. (The US Constitution stipulates that each state must have at least one representative, which would protect Carofornia in this case – the size of the House would have to be increased by one seat.)

Such quirks mean that seats in proportional systems are now generally apportioned using algorithms known as divisor methods. These work by dividing voting populations by a common factor so that when the fair proportions are rounded to a whole number they add up to the number of available seats. But this method is not foolproof: it sometimes gives a constituency more seats than the whole number closest to its fair proportion.

The balance of power

One criticism of proportional voting systems is that they make it less likely that one party wins a majority of the seats available, thus increasing the power of smaller parties as 'king-makers' who can swing the balance between rival parties as they see fit. The same can happen in a plurality system if the electoral arithmetic delivers a hung parliament, in which no party has an overall majority – as happened, for example, in the UK general election of 2017.

Where does the power reside in such situations? One way to quantify that question is through the Banzhaf power index. This involves, first, listing all combinations of parties that could form a majority coalition, and in all of those coalitions counting how many times a party is a 'swing' partner that could destroy the majority if it dropped out. Dividing this number by the total number of swing partners in all possible majority coalitions gives a party's power index.

For example, imagine a parliament of six seats in which party A has three seats, party B has two and party C has one. There are three ways to make a coalition with a majority of at least four votes: AB, AC and ABC. In the first two instances, both partners are swing partners. In the third instance, only A is – if either B or C dropped out, the remaining coalition would still have a majority. Among the total of five swing partners in the three coalitions, A crops up three times and B and C once each. So A has a power index of 3 ÷ 5, or 0.6, or 60 per cent – more than the 50 per cent of the seats it holds – and B and C are each 'worth' just 20 per cent.

9
Numbers and reality

Mathematics is the language of physics and all the sciences that build upon it. Here, we take up aspects of numbers we introduced earlier to explore in more detail an essential question touched upon at the start: to what extent do we live in a mathematical universe?

Is everything made of numbers?

When Albert Einstein (see Figure 9.1) finally completed his general theory of relativity in 1915, he looked down at the equations and discovered an unexpected message: the universe is expanding. But Einstein did not believe that the physical universe could shrink or grow, so he ignored what the equations were telling him. A decade or so later, Edwin Hubble and others found clear evidence of the universe's expansion. Einstein had missed the opportunity to make the most dramatic scientific prediction in history.

FIGURE 9.1 Albert Einstein was a physics genius – but how did his equations 'know' that the universe was expanding when no person did at the time?

How did Einstein's equations 'know' that the universe was expanding when he did not? If mathematics is just an invention of the human brain, how can it possibly churn out anything beyond what we put in? The prescience of mathematics seems no less miraculous today. At the Large Hadron Collider at CERN in Switzerland, physicists recently observed the fingerprints of a particle that was arguably discovered in 1964 by theorist Peter Higgs and others lurking in the equations of particle physics.

How is it possible that mathematics knows about Higgs particles or any other feature of physical reality? Maybe it's because it *is* reality, argues physicist Brian Greene of Columbia University in New York. Perhaps if we dig deep enough, we would find that physical objects like tables and chairs are ultimately not made of particles or strings, but of numbers. It might even be true to say that the whole universe is made of maths, not matter.

It is difficult, indeed, to understand what it means if we say that the universe is 'made of mathematics'. An obvious starting point is to ask what mathematics is made of. The late physicist John Wheeler said that 'the basis of all mathematics is $0 = 0$', referring to the idea that all mathematical structures can be derived from something called 'the empty set', the set that contains no elements (see Chapter 2). Keep nesting the nothingness like invisible Russian dolls and eventually all of mathematics appears.

That may be the ultimate clue to existence – after all, a universe made of nothing does not require an explanation. Indeed, mathematical structures do not seem to require a physical origin at all. According to physicist Max Tegmark of the Massachusetts Institute of Technology in Cambridge, USA, no dodecahedron has ever been created. To be created, something first has to not exist in space or time and then exist. A dodecahedron does not exist in space or time at all, he says — it exists independently of them. In turn, space and time themselves are contained within larger mathematical structures that just exist; they cannot be created or destroyed.

That raises a big question: why is the universe made of only some of the available mathematics? Today, only a tiny sliver of mathematics has a realization in the physical world. Pull any textbook off the shelf and most of the equations in it do not correspond to any physical object or physical process.

It is true that seemingly arcane and unphysical mathematics does, sometimes, turn out to correspond to the real world. For instance, the imaginary numbers were once considered totally deserving of their name, but are now used to describe the behaviour of elementary particles (see Chapter 5); non-Euclidean geometry eventually showed up as gravity. Even so, these phenomena represent a tiny slice of all the mathematics out there.

What about the mathematics our universe doesn't use? Tegmark thinks that other mathematical structures correspond to other universes. He calls this the 'level 4 multiverse', and it is far stranger than the multiverses that cosmologists like him often discuss. Their common-or-garden multiverses are governed by the same basic mathematical rules as our universe, but Tegmark's level 4 multiverse operates with completely different mathematics.

All of this sounds bizarre, but the hypothesis that physical reality is fundamentally mathematical has passed every test. If physics were to hit a roadblock so that it became impossible to proceed, we might find that nature cannot be captured mathematically. But, to Tegmark, it is really remarkable that that hasn't happened. Galileo said that the book of nature was written in the language of mathematics — and that was 400 years ago.

If reality is not, at bottom, mathematics, what is it? 'Maybe someday we'll encounter an alien civilization and we'll show them what we've discovered about the universe,' Greene says. 'They'll say, "Ah, math. We tried that. It only takes you so far. Here's the real thing." What would that be? It's hard to imagine. Our understanding of fundamental reality is at an early stage.'

Is the universe infinite?

Nowhere is the conflict between a mathematical conception of reality and reality itself more apparent than with infinity. Infinity is essential to the structure of mathematics (see Chapter 3). But when it comes to theories that aim to describe physical reality, its usefulness is less apparent. In much of physics, the appearance of an infinity generally means that there is something wrong with your theory.

Take the standard model of particle physics. While Einstein's general relativity explains gravity, the standard model explains all the other forces of nature. It is based on quantum theories and was long beset by pathological infinities. Quantum electrodynamics, the part of the standard model that deals with the electromagnetic force, initially showed the mass and charge of an electron to be infinite.

Decades of work, rewarded by many a Nobel Prize, banished these nonsensical infinities – or most of them. But gravity has notoriously resisted unification with the other forces of nature within the standard model because it seems immune to physicists' best tricks for neutralizing infinity's effects. In extreme circumstances, such as in a black hole's belly, Einstein's equations of general relativity break down as matter becomes infinitely dense and hot, and space-time infinitely warped.

But it is at the Big Bang that infinity wreaks the most havoc. According to the theory of cosmic inflation, the universe underwent a burst of rapid expansion in its first fraction of a second. Inflation explains essential features of the universe, including the existence of stars and galaxies. But the theory says it cannot be stopped. It continues inflating other bits of space-time long after our universe has settled down, creating an infinite multiverse in an eternal stream of Big Bangs. In an infinite multiverse, everything that can happen will happen an infinite number of times. Such a cosmology predicts everything – which is to say, nothing.

This disaster is known as the measure problem, because most cosmologists believe it will be fixed with the right 'probability measure' that would tell us how likely we are to end up in a particular sort of universe and so restore our predictive powers. But others think there is something more fundamental amiss. 'Inflation is saying, hey, there's something totally screwed up with what we're doing,' says physicist Max Tegmark. 'There's something very basic we've assumed that's just wrong.'

For Tegmark, that something is infinity. Physicists treat space-time as an infinitely stretchable mathematical continuum; like the line of real numbers, it has no gaps. Abandon that assumption and the whole cosmic story changes. Inflation will stretch space-time only until it snaps. Inflation is then forced to end, leaving a large, but finite, multiverse. Tegmark thinks all our problems with inflation and the measure problem stem directly from our assumption of the infinite. It is, he says, the ultimate untested assumption.

And there are good reasons to think we don't actually need infinity to describe the universe. Studies of the quantum properties of black holes by Stephen Hawking and Jacob Bekenstein in the 1970s led to the development of the holographic principle, which makes the maximum amount of information that can fit into any volume of space-time proportional to roughly one-quarter the area of its horizon. The largest number of informational bits a universe of our size can hold is about 10^{122}. If the universe is indeed governed by the holographic principle, there is simply not enough room for infinity.

Whether that is true or not, even the best device will never measure anything physical with infinite accuracy. The best atomic clocks measure increments of time out to fewer than 20 decimal places. The electron's anomalous magnetic moment, a measure of tiny quantum effects on the particle's spin, has been pinned down to 14 decimal places.

Tegmark, for his part, is intrigued by the fact that the calculations and simulations that physicists use to check a theory against the hard facts of the world can all be done on a finite computer. This shows that we don't need the infinite for the things we're doing. What's more, he points out, there's absolutely no evidence that nature is doing it any differently, that nature needs to process an infinite amount of information.

Quantum infinities

Seth Lloyd, a physicist and quantum information expert also at MIT, counsels caution with such analogies between the cosmos and an ordinary, finite computer. We have no evidence that the universe behaves as if it were a classical computer, he says. But there's plenty of evidence that it behaves like a quantum computer.

At first glance, that would seem to be no problem for those wishing to banish infinity. Quantum physics was born when, at the turn of the twentieth century, physicist Max Planck showed how to deal with another nonsensical infinity. Classical theories were indicating that the amount of energy emitted by a perfectly absorbing and radiating body should be infinite, which clearly was not the case. Planck solved the problem by suggesting that energy comes not as an infinitely divisible continuum, but in discrete chunks – quanta.

The difficulties start with Schrödinger's cat. When no one is watching, the famous quantum feline can be both dead and alive at the same time: it hovers in a 'superposition' of multiple, mutually exclusive states that blend together continuously. Mathematically, this continuum can only be depicted using infinities. The same is true of a quantum computer's 'qubits', which can perform vast numbers of mutually exclusive calculations simultaneously, just as long as no one is demanding an output. According to Lloyd, if you wanted to specify the full state of one qubit, it would require an infinite amount of information.

Max Tegmark is unfazed. He points out that when quantum mechanics was discovered, we realized that classical mechanics was just an approximation. He thinks another revolution is going to take place, and we shall see that continuous quantum mechanics is itself just an approximation to some deeper theory, which is totally finite.

For physicists looking for a way forward, it is easy to see the appeal of banishing infinity. Perhaps we might see the way to unify physics, uniting gravity with quantum theory. For Tegmark's particular concern, the measure problem, we would be freed from the need to find an arbitrary probability measure to restore cosmology's predictive power. In a finite multiverse, we could just count the possibilities. If there really were a largest number, then we would only have to count so high.

Mathematician Hugh Woodin of Harvard University would rather separate the two issues of physical and mathematical infinities, given how essential infinity has proved to be in set theory, the underpinning of modern mathematics. 'It may well be that physics is completely finite,' he says. 'But in that case, our conception of set theory represents the discovery of a truth that is somehow far beyond the physical universe.'

Is the universe random?

'Oh, I am fortune's fool,' says Romeo. Having killed Tybalt and realizing that he must leave Verona or risk death, he was expressing a view common in Shakespeare's time: that we are all marionettes, with some higher cause pulling the strings. We looked at randomness, chance and probability in Chapter 6. They might be fine mathematical tools for modelling a world where we know little, but is anything in the world truly random?

Long before dice were used for gaming, they were used for divination. Ancient thinkers thought the gods determined the outcome of a die roll; the apparent randomness resulted from our ignorance of divine intentions. Oddly, modern science at first did little to change that view. Isaac Newton devised laws of motion and gravitation that connected everything in the cosmos with a mechanism run by a heavenly hand. The motion of the stars and planets followed the same strict laws as a cart pulled by a donkey. In this clockwork universe, every effect had a traceable cause.

If Newton's universe left little room for randomness, it did at least provide tools to second-guess the Almighty's intentions. If you had all the relevant facts pertaining to a die roll at your fingertips – trajectory, speed, roughness of the surface and so on – you could, in theory, use mathematics to calculate which face would end up on top. In practice, this is far too complex a task for our brains or any computer we have yet invented. But it showed that randomness was nothing intrinsic; it was just a reflection of our lack of information (Figure 9.2).

Confidence in cosmic predictability led the French mathematician and physicist Pierre-Simon de Laplace to assert, a century after Newton, that a sufficiently informed intelligence could forecast everything that is going to happen in the universe – and, working backwards, tell you everything that did happen, right back to the cosmic beginnings. It's a glorious and rather discomfiting idea. If everything really is predictable, then surely all is pre-determined and free will is an illusion? Romeo, in other words, is right.

It was not until about two centuries after Newton that anyone began seriously to challenge the notion of a predictable cosmos. In 1859 the Scottish physicist James Clerk Maxwell

FIGURE 9.2 The way the world works often seems random – but that might just be because we lack the information to say why things happen as they do.

drew attention to the huge disparities in outcome that can stem from tiny factors affecting the collisions of molecules. This was the beginning of chaos theory. In its most familiar guise of the butterfly effect – that the flap of a butterfly's wings in Brazil might set off a tornado in Texas, as the chaos theorist Edward Lorenz put it in 1972 – this seems to restore unpredictability to the world. With a sufficiently complex system, even the tiniest approximation while working at the limits of your clock, barometer or ruler, or the slightest rounding error in a computation, can drastically affect the result. This is what makes the weather so hard to predict. Its eventual state is highly dependent on the initial measurement – and we can never have a perfect initial measurement.

As yet, we are only scratching the surface, however. While we seem to occupy a reality where causes lead to predictable effects, dig down and that is apparently not how things work at all. The advent of quantum theory brought us face to face with a rather different mathematical reality.

Quantum mathematics

Quantum theory is our working theory of reality at its most basic, explaining the workings of the fundamental particles of matter and the forces that act on them, with the notable exception of gravity. Developed in stages since the early twentieth century, it does away with cast-iron certainty entirely. Quantum experiments show us that nature is fundamentally random.

Fire a single photon of light at a half-silvered mirror, and it might pass through or be reflected: quantum rules give us no way to tell beforehand. Give an electron a choice of two slits in a wall to pass through, and it chooses at random. Wait for a single radioactive atom to emit a particle, and you might wait a millisecond or a century. This rather lackadaisical attitude to classical certainties could even account for why we are here in the first place. A quantum vacuum containing nothing can randomly and spontaneously generate something. Such a careless energy fluctuation might best explain how our universe began.

Explaining the explanation is trickier. We don't know where the quantum rules came from; all we know is that the mathematics behind them, rooted in uncertainty, corresponds to reality observed up close. That starts with the Schrödinger equation, which describes how a quantum particle's properties evolve over time. An electron's position, for example, is given by an 'amplitude' smeared over space contained within a probabilistic mathematical wave function that describes all the possible states or places it might be in. There is a set of mathematical rules you can apply to the wave function to find the probability that any particular measurement will pinpoint the electron to any particular position.

This does not guarantee that the electron will be in that position at any one time. But by repeatedly doing the same measurement, resetting the system each time, the distribution of results will match the Schrödinger equation's predictions. The repeated, predictable patterns of the classical world are ultimately the result of many unpredictable processes.

The repercussions are interesting. Say you want to walk through a wall; the mathematics of quantum theory says that it is possible. Each one of your atoms has a position that could – randomly – turn out to be on the other side of the wall when it interacts. That event's probability is exceedingly low, and the probability that all of your atoms will simultaneously locate to the other side of the wall is infinitesimally small. A nasty bruise is the sum of all the other probabilities. Welcome to reality.

Which probability is right?

Quantum randomness throws up other conundrums. The idea that an objective, universally valid view of the world can be achieved by making properly controlled measurements is perhaps the most basic assumption of modern science. It works well enough in the macroscopic, classical world. Kick a football, and Newton's laws of motion tell you where it will be later, regardless of who is watching it and how.

Kick a quantum particle such as an electron or a quark, though, and the certainty vanishes. At best, quantum theory allows you to calculate the probability of one outcome from many encoded in the multifaceted wave function that describes the particle's state. Another observer making an identical measurement on an identical particle might measure something very different. You have no way of saying for sure what will happen.

So here's the question: what state is a quantum object in when no one is looking? The most widely accepted answer comes in the form of the Copenhagen interpretation, named after the home of a school of early quantum theorists centred around the pioneer Niels Bohr. Schrödinger's notorious cat, devised in a thought experiment by the Austrian physicist Erwin Schrödinger in 1935, illustrates its conclusion. Shut in a box with a vial of lethal gas that might, or might not, have been released by a random quantum event such as a radioactive decay, the unfortunate feline hangs in limbo, both alive and dead. Only when an observer opens the box does the cat's wave function 'collapse' from its multiple possible states into a single real one.

This opens a physical and philosophical can of worms. Einstein pointedly asked whether the observations of a mouse would be sufficient to collapse a wave function. If not, what is so special about human consciousness? If our measurements truly do affect reality, this also opens the door to effects such as 'spooky action at a distance' – Einstein's dismissive phrase to describe how observing a wave function can seemingly collapse another one simultaneously on the other side of the universe.

Then there is the mystery of how atoms and particles can apparently adopt split personalities, but macroscopic objects such as cats clearly cannot, despite being made up of those same atoms and particles. Schrödinger's intention in introducing his cat was to highlight this inexplicable division between the quantum and classical worlds. The split is not only there, but it is also 'shifty', as quantum theorist John Bell described it: physicists contrive to put ever-larger objects into fuzzy quantum states, for instance, so we have no set way of defining where the boundary lies.

The Copenhagen interpretation simply ignores these quantum mysteries, saying that we should just accept that the mathematics works and have done with it. In 1989 physicist David Mermin of Cornell University dubbed it the 'shut up and calculate' approach, a name that has stuck among detractors.

There are alternative physical interpretations. A prominent one is the many worlds interpretation, which suggests that the universe divides into different paths every time anything is observed. But none quite seems to crack the central mystery.

One possibility that some physicists, Mermin included, have recently begun to champion is that mystery really does lie in the maths – in fact, in our interpretation of the wave function probabilities that seem to govern the quantum world. The idea is known as quantum Bayesianism, after one of the two main strands of thinking in probability theory (see Chapter 6). Conventionally, quantum probabilities are viewed as frequentist probabilities. In the same way that you might count up many instances of a coin falling heads or tails to conclude that the odds are 50/50, many measurements of a quantum system tell you the relative frequency of its multiple states cropping up.

Despite its limitations, not least when dealing with single, isolated events, frequentist probability is popular throughout science for the way it turns an observer into an entirely objective counting machine. Bayesian probability, on the other hand, allows you to acquire a new piece of information and update your assessment of the likelihood of something happening (see Figure 9.3).

Standard quantum picture

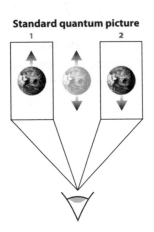

Objects in the quantum world exist in a fuzzy combination of states. The act of measuring forces them to adopt a specific state (1 or 2).

Quantum Bayesianism

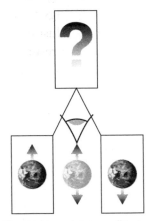

The quantum states are all in our minds – they are just a fluid tool we use to understand our variable experiences of the world.

FIGURE 9.3 Quantum Bayesianism is an alternative to the standard picture of the quantum world. It uses Bayesian statistics to say its apparent uncertainty is all in our minds.

The central argument of quantum Bayesianism, or QBism, is that this more subjective type of probability applies in the quantum world. Measure the position of an invisible electron, say, and you acquire new knowledge, and update your assessment of the probabilities accordingly, from uncertain to certain. Nothing needs to have changed at the quantum level. Quantum states, wave functions and all the other probabilistic apparatus of quantum mechanics do not represent objective truths about stuff in the real world. Instead, they are subjective tools that we use to organize our uncertainty about a measurement before we perform it. In other words, quantum weirdness is all in the mind.

For Mermin, the beauty of the idea is that the paradoxes that plague quantum mechanics simply vanish. Measurements do not 'cause' things to happen in the real world, whatever that is; they cause things to happen in our heads. Spooky action at a distance is an illusion, too. The appearance of a spontaneous change is just the result of two parties independently performing measurements that update their state of knowledge.

As for that shifty split, the 'classical' world is where acts of measurements are continuous, because we see things with our own eyes. The microscopic 'quantum' world, meanwhile, is where we need an explicit act of measurement with an appropriate piece of equipment to gain information. To predict outcomes in this instance, we require a theory that can take account of all the things that might be going on when we are not looking. For a QBist, the quantum/classical boundary is the split between what is going on in the real world and your subjective experience of it.

But does that really solve any mystery? Critics of QBism say that it depends on the idea that there is no direct experience of things, only that which we construct in our minds from sensory inputs. Imagine, for example, setting up some equipment to measure the energy of a particle, and then going off for a cup of tea. During the tea break, did the pointer on the equipment's dial have no definite orientation? A QBist would say maybe not, you can't tell – even though experience tells us that a macroscopic object such as a pointer does always have a definite orientation.

Many physicists are less than happy with this result. Carlo Rovelli of Aix-Marseille University in France says he would prefer an interpretation of quantum theory that made sense even if there were no humans to observe anything. Caslav Brukner of the University of Vienna in Austria sees a limitation in the fact that QBism seems to lack the power to explain why quantum theory has the mathematical and conceptual structure it does. However you interpret this particular mathematical construction, the mystery of why it works in describing and predicting the outcome of particle experiments remains.

Can mathematics reveal a theory of everything?

The two big breakthroughs in physics in the twentieth century owed much to mathematics. Although it tells us that nature is intrinsically fuzzy, the mathematics of quantum theory also explains how atoms behave in precise, predictable ways when they emit and absorb light, or link together to make molecules. And it pointed the way to new discoveries, most famously leading the British theorist Paul Dirac to formulate an equation that led to the idea of antimatter several years before the first antimatter was found in 1932.

The second big breakthrough was Einstein's general relativity. More than 200 years earlier, Isaac Newton had shown that the force that makes apples fall is the same as the gravity that holds planets in their orbits. Newton's mathematics is good enough to fly rockets into space and steer probes around planets, but Einstein transcended Newton. His general theory of relativity could cope with very high speeds and strong gravity, offering deeper insight into gravity's nature.

Einstein was not a top-rate mathematician, and he was lucky that the geometrical concepts he needed for general relativity had already been developed by the German mathematician Bernhard Riemann a century earlier (see Chapter 7). The young quantum theorists were also able to apply ready-made mathematics. Today's physicists are not so lucky.

The big challenge in fundamental physics now is to mesh general relativity and quantum mechanics into a unified 'theory of everything'. The most favoured, although by no means the only, approach is string theory, which has at its core the idea that the particles that make up atoms are all made up of tiny loops, or strings, that vibrate in a space with 10 or 11 dimensions.

String theory, if correct, would vindicate the vision of Einstein and others that the world is essentially a geometrical structure. But it involves intensely complex mathematics that certainly cannot be found on the shelf, and it is as yet far from providing convincing answers. Arguments rage over whether string theory is right, whether it will ever engage with experiment, and even whether it is physics at all (see Figure 9.4).

If we fail to make progress, it may be because, although a 'true' fundamental theory exists, it is just too hard for human brains to grasp. A fish may be barely aware of the medium in which it lives and swims; certainly it cannot understand that water consists of interlinked atoms of hydrogen and oxygen. Space and time form the medium in which we live; it could be that their microstructure is, likewise, too complex for our mathematical ability.

And if, as many lines of thinking in physics seem to indicate, our universe is just one in a multiverse, other branches of

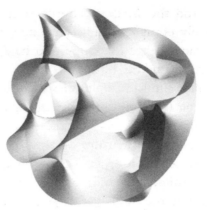

FIGURE 9.4 String theory, our best hope for a theory of everything, involves compact extra dimensions shaped according to an abstruse mathematical geometry known as a Calabi–Yau manifold.

mathematics might then become relevant. We need a rigorous language to describe the number of possible states that a universe could possess and to compare the probability of different configurations. A clearer concept of infinity would also be required (see Chapter 3).

Provided we could understand it, a unified theory would be an intellectual triumph. Calling it a 'theory of everything', though, is misleading. String theory unifies the very big and the very small, but there is a third frontier – the very complex.

Simple underlying rules might govern some seemingly complex phenomena, too. This was intimated in 1970 by the mathematician John Conway, who invented the 'game of life'. Conway wanted to devise a game that would start with a simple pattern and use basic rules to evolve it again and again. He began experimenting with the black-and-white tiles on a Go board and discovered that by adjusting the simple rules of his game, which determine when a tile turns from black to white and vice versa, and the starting patterns, some arrangements produce incredibly complex results seemingly from nowhere. Some patterns can emerge that appear to have a life of their own as they move round the board.

The real world is similar: simple rules allow complex consequences. Exploring these consequences is less a case of better theories and more complex mathematics, and more a case of better ways to use the mathematics we already have.

While Conway needed only a pencil and paper to devise his game, it takes a computer to fully explore the range of complexity inherent in it. Increased computing power and the ability of machines to learn mean that we are at the point where computers, working on the logic of existing mathematics, may have the capacity to solve problems that have so far eluded us: high-temperature superconductors, perhaps, or how combinations of genes encode the intricate chemistry of a cell. Even so, there are likely to be limits to what they can tell us.

The limits to computation

If we were able mathematically to simulate the fine-grained movements of all the universe's matter, we might be able to predict its evolution and its fate. There is just one problem: with current computing power, it would take more time than the universe has to offer. Computational power is a practical limitation we can blame for everything, from unreliable weather forecasts to shoddy logistics. As we have already seen in Chapters 7 and 8, problems such as optimizing an itinerary rapidly become too hard to compute when there are more than a few thousand destinations involved.

But, ultimately, blaming computers is just a fig leaf for a mega-sized limitation. However powerful we make them, computers ultimately rely on human input to program them – and human thought is a mess. Statements like 'this statement is false', or actions like hating someone yet loving them, both compute and do not compute. 'Language is an expression of the mind, and my mind and language are full of contradictions,' says Noson Yanofsky, an information scientist at the City University of New York.

Logic, and the mathematics that builds on it, is supposedly our way out: a cleaner, neutral language for a trained brain to describe in abstract terms what it cannot visualize. That is all very well – but then there are the logical limitations of mathematics itself. These start with well-known injunctions such as never to divide a number by zero (see Chapter 2). If maths is the language of a flawless universe, we cannot allow the logical inconsistencies that result from taking such a step – so we don't. 'If you want mathematics to continue without contradictions, then you have to somehow restrict yourself,' says Yanofsky.

And as Kurt Gödel showed with his incompleteness theorems of the 1930s (see Chapter 3), any system of logic containing the rules of arithmetic is bound to contain statements that can be neither proved nor disproved. Gödel incompleteness is a mathematical expression of the logical–illogical statement 'this statement is false'. In fact, there is no way for anything, be it a simple sentence, a system of logic or a human being, to express the full truth about itself.

This problem of self-reference is endemic. Gödel's contemporary Alan Turing showed that you cannot ask a computer program in advance whether it will run successfully. Quantum mechanics sprouts paradoxes because we are part of the universe we are trying to measure. And whether mathematics is invented or discovered, whether it is all there is or just a tool to explain the world, the frustrating yet exhilarating conclusion is that we shall never fully get to grips with it – because we are the ones doing it.

Conclusion

At the end of our journey through the world of numbers, mathematician Ian Stewart sums up what it all means – and where our mathematical voyage might end.

What makes maths special?

The ancient Greek philosopher Plato held that mathematical concepts actually exist in a strange kind of ideal reality just off the edge of the universe. A circle is not just an idea; it is an ideal. We imperfect creatures may aspire to that ideal, but we can never achieve it, if only because pencil points are too thick for us to draw it. But there are those who say that mathematics exists only in the mind of the beholder. It does not have any existence independent of human thought, any more than language, music or the rules of football do.

Who is right? There is much that is attractive in the Platonist point of view. It is tempting to see our everyday world as a pale shadow of a more perfect, ordered, mathematically exact one. Mathematical patterns show up throughout the world and they have a universal feel to them. The catalogue of patterns is also remarkably versatile, with the same patterns being used in many different guises. Raindrops and planets are both spherical. Rainbows and ripples on a pond are circular. Honeycomb patterns are used by bees to store honey, but can also be found in the geographical distribution of territorial fish, the frozen magma of the Giant's Causeway, and rock piles created by convection currents in shallow lakes. Spirals can be seen in water running out of a bath and in the Andromeda Galaxy.

With this kind of ubiquitous occurrence of the same mathematical patterns, it is no wonder that physical scientists get carried away and declare them to lie at the very basis of space, time and matter. Physicist Eugene Wigner called it the 'unreasonable effectiveness' of mathematics. Long before him, Plato wrote that 'God ever geometrizes'. The physicist James Jeans declared that God was a mathematician. Paul Dirac thought that He was a pure mathematician. One stream of thought prevalent in physics today holds that reality is made from information, the raw material of mathematics.

These are powerful, heady ideas, and they are highly appealing to mathematicians. However, it is equally conceivable that all of this apparently fundamental mathematics is in the eye of the beholder or, more accurately, in the beholder's mind. We human beings do not experience the universe raw but through our senses, and we interpret the results using our minds. So to what extent are we mentally selecting particular kinds of experience and deeming them to be important, rather than picking up things that really are important in the workings of the universe?

The answer, I think, is a bit of both. When we are trying to formalize an elusive idea or find a new method, it feels like invention: we are floundering around, trying all sorts of ideas, and we simply do not know where it will all lead. But the more established an area of mathematics becomes, the more strongly it feels as if there is some kind of fixed logical landscape we are merely exploring. Once we have made a few assumptions in the form of axioms, then everything that follows from them is predetermined.

But if mathematics just resides in mathematicians' heads, why is it so 'unreasonably effective'? The easy answer is that most mathematics starts in the real world. For instance, after observing on innumerable occasions that two sheep plus two more sheep make four sheep, ditto cows, wolves, warts and witches, it is a small step to introduce the idea that $2 + 2 = 4$ in a universal, abstract sense. Since the abstraction came out of reality, it is no surprise if it applies to reality.

However, that is too simple-minded a view. Mathematics has an internal structure of logical deduction that allows it to grow in unexpected ways. New ideas can be generated internally, too, whenever anyone tries to fill obvious holes in the logical landscape. For example, having worked out how to solve quadratic equations – those involving squared quantities, which turn up in all sorts of practical situations – it is obvious that you ought to try to solve cubic, quartic and quintic equations, too. Before you can say 'Evariste Galois' you're doing Galois theory, which shows that you cannot solve quintics, but is almost totally useless for anything practical. Then someone generalizes Galois theory so that it applies to differential equations, and suddenly you find applications again, in a completely different area.

Yes, there is a flow of problems and concepts from the real world into mathematics, and a back-flow of solutions from mathematics to reality. Wigner's point is that the back-flow may not answer the problem that you set out to solve. Why should this be? It is because mathematics is the art of drawing necessary conclusions independently of interpretations. Two plus two has to be four, whether you are discussing sheep, cows or witches. In other words, the same abstract structure can have several interpretations. You can get the ideas from one interpretation and transfer the result to others. Mathematics is so powerful because it is an abstraction.

This is all very well, but why do the abstractions of mathematics match reality? Indeed, do they really match, or is it all an illusion? Science, as opposed to mathematics, has an external reality check: theories must accord with observations. If the world's scientists all got together and decided that elephants are weightless and rise into the air if they are not held down by ropes, it would still be foolish to stand under a cliff when a herd of elephants was leaping off the edge. The reality check cannot be perfect because it is done by beings who see reality through imperfect and biased senses, but science still has to survive some very stringent scrutiny.

What is the reality check in maths? The deeper we delve into the 'fundamental' nature of the universe, the more mathematical it seems to get. The ghostly world of the quantum cannot be expressed without mathematics (see Chapter 9): if you try to describe it in everyday language, it makes no sense. Of course, not all fields are so obviously mathematical in their structure. The biological world, in particular, seems not to obey the rigid rules that we find in physics. The 'Harvard law of animal behaviour' – that in carefully controlled laboratory conditions animals do just what they please – is more appropriate than Newton's laws of motion.

But the problem here could be a difference of scale. Quantum physics tends to be applied to simple arrangements of matter – a few atoms, say. In biology, the significant arrangements of matter are enormously more complex: there are trillions of atoms in the human genome, and this is just one DNA strand inside one cell of a much more complex organism. An atom-by-atom description of a human being would involve numbers with a huge number of zeros. Human beings could well be behaving according to mathematical rules – but it is mathematics so complicated that human mathematicians cannot possibly write it down, let alone grasp what it means. Moreover, it is mathematics whose structure is almost totally impenetrable, for the boring reason that there is just too much information to take in.

This is the old philosophical problem of 'emergence', but in a new guise. Emergent phenomena are things that seem to transcend their ingredients, like consciousness arising in a material brain. Your behaviour is caused by mathematical rules applied to your constituent atoms, in the context of everything that is happening around you, but you cannot do the calculations to check that because they are too messy and too lengthy.

You could argue that this makes the whole question academic: it does not matter whether this kind of mathematical basis exists for biology because, even if it does exist, it is of no practical use. However, there is an attractive alternative. Even very complex mathematical systems tend to generate recognizable patterns on higher levels of description. For example, the underlying quantum theory of a crystal involves just as many atoms as a human being – at least if it's a human-sized crystal – and therefore runs into the same intractable problem of emergence.

But crystals exhibit clear mathematical patterns of their own, such as a regular geometric form, and while nobody can deduce this in full logical rigour from the quantum mechanics of their atoms, there is a chain of reasoning that makes it plausible that the laws of quantum mechanics do indeed lead to the regularities of crystal structure.

Roughly speaking, it goes like this: quantum mechanics causes the atoms to arrange themselves in a minimum-energy configuration; the overall symmetry of the laws of nature in space and time causes such configurations to be highly symmetrical; in this case, the consequence is that they form regular atomic lattices.

From this point of view, mathematical patterns that arise in high-level descriptions of living organisms are evidence that biology, too, is mathematical at heart. One example might be the appearance of Fibonacci numbers in the petals of flowers (see Chapter 5).

But do patterns like these really tell us that mathematics is inherent in nature? Our minds certainly have a tendency to seek out mathematical patterns, whether or not they are actually significant. This tendency has led to Newton's law of gravity and the equations of quantum mechanics, and also to astrology and an obsession with the measurements of the Great Pyramid. Ironically, what mathematics tells us about choosing lottery numbers is that any patterns we think we see are illusions (see Chapter 8).

It is worth asking how our minds developed this tendency for pattern seeking. Our human minds evolved in the real world, and they learned to detect patterns to help us survive events outside ourselves. If none of the patterns detected by these minds bore any genuine relation to the real world out-side, they would not have helped their owners survive and we would eventually have died out. So our figments must corre-spond, to some extent, to real patterns.

In the same way, mathematics is our way of understanding certain features of nature. It is a construct of the human mind, but we are part of nature, made from the same kind of matter, existing in the same kinds of space and time as the rest of the uni-verse. So the figments in our heads are not arbitrary inventions. There are definitely some mathematical things in the universe, the most obvious being the mind of a mathematician. Math-ematical minds cannot evolve in an unmathematical universe.

But that is not to say that only one kind of mathematics is possible: the mathematics of the universe. That seems too parochial a view. Would aliens necessarily come up with the same kind of mathematics as us? I don't mean in fine detail. For example, the six-clawed cat creatures of Apellobetnees Gamma would no doubt use base-24 notation, but they would still agree that 25 is a perfect square, even if they write it as 11.

However, I'm thinking more of the kind of mathematics that might be developed by the plasma-vortex wizards of Cygnus V, for whom everything is in constant flux. I bet they would understand plasma dynamics a lot better than we do, although I suspect that we would not have any idea how they did it. But I doubt that they would have anything like Pythagoras's theo-rem. There are few right angles in plasmas. In fact, I doubt they would have the concept 'triangle'. By the time they had drawn the third vertex of a right triangle, the other two would be long gone, wafted away on the plasma winds. So that is perhaps the real reason for the unreasonable effectiveness of mathematics: it is not that it exists in some Platonic realm; it is that we invent, or discover, the mathematics that fits reality around us.

Forty-nine ideas

This section offers 7^2 extra ideas for how to explore the world of numbers in greater depth.

Seven places of mathematical pilgrimage

1 **The National Museum of Mathematics in New York**, located on East 26th Street between 5th and Madison Avenues, bills itself as the only museum dedicated to maths – or math – in North America. Its exhibits include a square-wheeled tricycle that rolls smoothly on a specially designed surface.

2 **The Winton Gallery at London's Science Museum**, which opened in 2016, is dedicated to maths. Its swirly ceiling, designed by the star architect Zaha Hadid, represents the mathematical equations that describe air flow.

3 **The Seven Bridges of Königsberg.** Only one of these still stands – most of the city, now the Russian enclave of Kaliningrad sandwiched between Lithuania and Poland on the Baltic Sea, was destroyed during the Second World War. Leonhard Euler's proof that you could not cross all seven just once without backtracking is widely considered to be the founding of the discipline of graph theory. The word is that the present bridge configuration makes it possible: perhaps it is worth a walking tour on the Baltic to find out.

4 **The old city of Syracuse in Sicily, Italy**, is a World Heritage Site and well worth a visit. For maths buffs, it was the scene of Archimedes' apocryphal 'Eureka!' moment when he realized the mathematical law of displacement in his bathtub. Equally unverified is the story that Archimedes met his end when a Roman soldier, part of an army that had just sacked Syracuse, came across him drawing diagrams in the sand – and, against the orders of his general, stabbed him.

5 **Broom Bridge, on the Royal Canal in Dublin's northern suburbs,** has a plaque with the equation for four-dimensional complex numbers, or quaternions – celebrating the location where mathematician William Rowan Hamilton was inspired to create them.

6 **Tilings at the Alhambra in Granada, Spain,** show complex periodic patterns – in fact, they contain examples of 13 of the 17 classes of periodic symmetry. For fans of less regularity, the terrace outside the Andrew Wiles Building of the Mathematical Institute in Oxford, UK, is paved with an aperiodic Penrose tiling.

7 **Göttingen, Germany,** was the most significant place in the development of modern mathematics. There is nothing of particular mathematical interest to see there today, but in the nineteenth and early twentieth century its university was home to a series of seminal mathematicians, among them Carl Friedrich Gauss, David Hilbert, Emmy Noether and Bernhard Riemann.

Seven quirky integers

1 **1** It may be, unlike 0, indisputably a number, but 1 has properties that make it stand out. Just as zero is the 'additive identity' – adding 0 to anything changes nothing – 1 is the 'multiplicative identity'. Any number multiplied by 1 doesn't change, including 1 itself. It follows that 1 is the only number that is its own square, its own cube and so on. It is also the only natural number that is neither a prime number (it is divisible only by 1, so falls down on that definition) nor a 'composite' number that can be produced by multiplying two smaller natural numbers together.

2 **6** In his mathematical primer *The Elements*, Euclid coined the term 'perfect number'. It refers to a number that is the sum of all the numbers it divides by, excluding itself. Thus $6 = 1 + 2 + 3$ is the first example; the next are 28, 496 and 8128.

3 **70** The quirkiness of this number speaks for itself: it is the smallest 'weird number'. A weird number has two qualities. First, it is an 'abundant' or 'excessive' number, meaning that the sum of all the numbers it divides by, including 1 but not including itself, is bigger than itself: in the case of 70, $1 + 2 + 5 + 7 + 10 + 14 + 35 = 74$. But a weird number is also not 'semi-perfect'; that is to say, no subset of those divisors adds up to the number itself. This is a rare combination – after 70, the next examples are 836 and 4030.

4 **1729** This number would have been seen as unremarkable were it not for an anecdote told by the mathematician G. H. Hardy about his friend and mentee Srinivasa Ramanujan, to illustrate the latter's peculiar brilliance.

Hardy wrote: 'I remember once going to see him when he was ill at Putney. I had ridden in taxi cab number 1729 and remarked that the number seemed to me rather a dull one, and that I hoped it was not an unfavourable omen. 'No,' he replied, 'it is a very interesting number; it is the smallest number expressible as the sum of two cubes in two different ways.' Those ways are $1^3 + 12^3$ and $9^3 + 10^3$, and 1729 has since been known as the Hardy-Ramanujan number.

5 **3435** This is a Münchhausen number, one of only two known to exist. Named after the German nobleman Baron von Münchhausen who was known for his tall tales, 3435 can 'raise itself' – it is the sum of its digits raised to their own power. $3^3 + 4^4 + 3^3 + 5^5 = 27 + 256 + 27 + 3125 = 3435$. The other Münchhausen number is, of course, 1.

6 **6174** Take any four-digit number that contains at least two different digits. Arrange the digits in descending and ascending order, and subtract the second from the first. Repeat the previous step with this new number (treating any zeros as normal digits), until the result of the subtraction is 6174. As the Indian mathematician D. L. Kaprekar noted in 1955, this will happen in no more than seven steps.

7 **Graham's number** In the 1970s mathematician Ronald Graham was working on a problem to do with cubes in higher dimensions. When he finally got there, the answer involved a number that was not infinite but so large that it could not be written down – there is literally not enough space in the universe. It follows that we cannot reproduce it here, although we do know that its last digit is 7.

Seven (seeming) paradoxes

1 **The coastline paradox** How long is the coast of Britain? That was the question mathematician Benoit Mandelbrot asked in 1969, in a paper with the subtitle 'Statistical self-similarity and fractional dimension'. In it, he explored an apparent paradox. It is that the length of a coastline of an island such as Great Britain, with its many inlets, curves and complications on all sorts of scales, depends on the length of the thing you use to measure it. The smaller the ruler, the longer it becomes as you take into account more and more ins and outs. Yet it clearly does have a set length, doesn't it?

   For Mandelbrot, the answer was that it makes no sense to treat a shape like a coastline with repeating complications on many scales as a straight line with a length in just one dimension. On the other hand, a coastline is clearly not a shape in two dimensions, either. It is somewhere in between – it has the counter-intuitive property of having a fractional dimension. In the word Mandelbrot introduced to describe such patterns in 1975, he kick-started a whole new avenue of mathematical discovery, 'fractals'.

2 **Zeno's paradox** Motion is impossible and change does not happen. That is the lesson of a series of paradoxes devised by the Greek philosopher Zeno of Elea in the fifth century BCE. The best known is that of Achilles and the tortoise. Achilles gives a tortoise a head start, say of 100 metres, in a race. After a certain time, Achilles has run 100 metres, in which time the tortoise has moved 10 metres, so is still ahead. But in the time it

takes Achilles to move that extra 10 metres, the tortoise has moved on again. In fact, you can show that, although Achilles will get very close to the tortoise, he will never actually overtake it.

Only in the late nineteenth century, with the application of some nifty calculus and a full appreciation of the mathematics of the infinite series that the problem represents, did we have what appears to be a mathematically watertight resolution of the paradox that conforms with experience: yes, Achilles can overtake the tortoise.

3 **The paradox of the heap** Also known as the Sorites paradox, this highlights the importance to any mathematical argument of precise, logically defined terms. You have a heap of grains of sand, which you remove one by one. Removing any one grain of sand does not turn the heap into a non-heap. But then again, a single grain of sand is not a heap. So if you keep on removing grains, when does it turn from a heap to a non-heap? It sounds trivial, but resolutions tend to require either setting an arbitrary numerical boundary of the size of a heap, denying that heaps can exist in the first place, or introducing a new three-valued logic that allows states of heap, non-heap and neither one nor the other.

4 **The elevator paradox** The cosmologist George Gamow had an office on the second floor of his building, while his colleague Marvin Stern was on the sixth floor, near the top. The two observed an odd fact: whereas the first elevator that came to Gamow's floor was almost always going down, the first to arrive at Stern's was almost invariably going up – even though (unless a continual

flow of elevators was originating somewhere in the middle of the building) just as many elevators should have been going up as down at each floor. This is, in fact, a real effect – but only some complex modelling of where an elevator spends most of its time in a building could explain why it should be the case.

5 **The friendship paradox** Most people have fewer friends than their friends have, on average. What appears to be a paradox is also a real effect, and has to do with the structure of the sort of networks that permeate our social circles. In simple terms, people with a large number of friends have a greater likelihood of being in your own group of friends, skewing the average. A similar effect means that, on average, most people's partners have had a greater number of other sexual partners.

6 **Simpson's paradox** In 1973 an analysis of admissions to the graduate school at the University of California, Berkeley, showed that men applying were more likely to be admitted than women. But when the people conducting the analysis broke it down by individual departments, there were more departments with a significant bias towards women. It is a famous example of Simpson's paradox, in which a trend seen in different groups of data disappears when those groups are combined. It is the curse of many a medical trial when researchers are trying to work out whether the effect of a drug is real across a whole population, say. In the UC Berkeley case, it turned out that the phenomenon was accounted for by more women applying to competitive departments with low general rates of admission than to any pervasive gender bias.

7 **Gabriel's horn paradox** By taking a graph of a particular mathematical function ($f(x) = 1/x$ for the domain
$x > 1$) and rotating it in three dimensions around the
x-axis, it is possible to create a shape that, although it
has an infinite surface area, has a finite volume. Known
as Gabriel's horn, or Torricelli's trumpet, the resulting
shape would be practically puzzling: it could, for example, hold only a finite amount of paint while requiring
an infinite amount of paint to coat its surface.

Seven lesser-known great mathematicians

1 **Muhammad ibn Mūsā al-Khwārizmī** (*c.*780–*c.*850) was a Persian mathematician whose writings, once translated into Latin, transformed Western mathematics, introducing among other things the decimal number system. The word 'algebra' derives from *al-jabr*, an operation he used to solve quadratic equations. The Latinized form of his name, Algoritmi, gives us the word 'algorithm'.

2 **Gerolamo Cardano** (1501–76) was a polymathic mathematician who, among other things, was the first to make use of imaginary numbers. He was also a compulsive gambler – a trait that led him to write the first systematic study of probability.

3 **Carl Friedrich Gauss** (1777–1855) is one of history's most influential mathematicians, making wide-ranging contributions to fields from number theory to statistics. He was an obsessive perfectionist and many of his results were not published in his lifetime. Today he is perhaps best known for the Gaussian or normal distribution in statistics, which predicts how a random quantity – such as the height of all mathematicians – will group around an average value.

4 **Evariste Galois** (1811–32) founded several branches of abstract algebra, as well as laying the groundwork for group theory. He did all this while being a radical adherent of French revolutionary ideals, dying in mysterious circumstances in a duel at the age of just 20.

5 **Emmy Noether** (1882–1935) was described by Albert Einstein as 'the most significant creative mathematical genius thus far produced since the higher education of women began'. Others have argued that only the first part of the sentence is necessary. The theorem that bears her name – that mathematical symmetries translate into conserved physical quantities – has provided a roadmap for discoveries in fundamental physics. Denied a full professorship in Göttingen because she was a woman, as a Jew she died in exile in the USA, a victim of Nazi racial laws.

6 **John von Neumann** (1903–57) is often called the last of the universal mathematicians. He made seminal contributions to game theory and the development of computing, as well as contributing to the development of quantum theory and the nuclear bomb. He was also famed for his photographic memory, sometimes entertaining friends by reciting pages from the telephone directory and entire works of literature such as *A Tale of Two Cities*.

7 **Paul Erdős** (1913–96) was a Hungarian-born mathematician known as 'the oddball's oddball'. He led an itinerant life travelling from conference to conference, eschewing most possessions and showing up at colleagues' houses announcing, 'My brain is open.' His uniquely collaborative style led him to co-author more than 1500 papers in his lifetime, and to the concept of the Erdős number as a measure of a mathematician's standing in the field. If you have an Erdős number of zero, you are Erdős himself; if you have a number one, you have collaborated with him; if you have number two, you have collaborated with a collaborator of his … and so on.

Seven maths jokes

... with seven maths-type explanations.

1 Why did the chicken cross the Möbius strip?
To get to the same side.

(A Möbius strip is a topological form with only one side.)

2 Two statisticians go hunting. The first one fires at a bird but overshoots by a foot. The second one fires and undershoots by a foot. They high-five and say 'Got it!'

(It's the law of averages ...)

3 Why do mathematicians like forests?
Because of all the natural logs.

(The natural log(arithm)s are those based on Euler's number, *e*.)

4 A: 'What is the integral of 1/cabin?'
B: 'Log cabin.'
A: 'No, houseboat – you forgot the C.'

(It's integral calculus – performing this operation on a function $1/x$ produces the (natural) logarithm of x as an answer – as long as you remember to add a 'constant of integration', C.)

5 A physicist, a biologist and a mathematician are sitting on a bench across from a house. They watch as two people go into the house, and then, a little later, three people walk out. The physicist says, 'The initial measurement was incorrect.' The biologist says, 'They must have reproduced.' And the mathematician says, 'If exactly one person enters that house, it will be empty.'

(Only a mathematician truly counts on negative numbers.)

6 Infinitely many mathematicians walk into a bar. The first says, 'I'll have a beer.' The second says, 'I'll have half a beer.' The third says, 'I'll have a quarter of a beer.' The barman pours just two beers. 'Is that all you're giving us?' the mathematicians ask. The bartender says: 'Come on, guys. Know your limits.'

(You can prove mathematically that the sum, or limit, of the infinite number of terms in the sequence $1/2^n$, so $1 + 1/2 + 1/4 \ldots$, is 2.)

7 What is a polar bear?
A Cartesian bear after a coordinate transformation.

(Cartesian and polar coordinates are two alternative coordinate systems.)

Seven maths films

1 *Good Will Hunting* **(1997)** is a fictional story of Will Hunting, a janitor at MIT with genius-level mathematical ability – but who, to thrive, must first overcome his demons.

2 **π (1998)** is a horror-thriller whose protagonist is a number theorist who sees mathematical patterns in everything around him, with unpalatable results.

3 *A Beautiful Mind* **(2001)** is a dramatized account of the life of John Nash (1928–2015), an American mathematician and game theory pioneer who won both the Abel Prize and the Nobel Prize for Economics, despite suffering from paranoid schizophrenia.

4 *Proof* **(2005)** is a fictional story based on a Pulitzer prize-winning play. It centres on a dispute about the ownership of a proof discovered in a deceased mathematician's effects. Cambridge mathematician and Fields Medallist Timothy Gowers acted as a consultant.

5 *Travelling Salesman* **(2012)** is an intellectual thriller based around four mathematicians who discover a solution to the notorious 'P = NP?' problem of computational complexity (see Chapter 7) and the moral consequences of their discovery.

6 *The Imitation Game* **(2014)** is a historical drama film centred around Alan Turing and other mathematicians who decrypted the Nazi Enigma codes during the Second World War.

7 *The Man Who Knew Infinity* (2015) tells the true story of the Indian mathematical genius Srinivasa Ramanujan (1887–1920) and his remarkable collaboration with the British mathematician G. H. Hardy.

Seven ideas for further reading

1 **The Very Short Introductions series** published by Oxford University Press are small books written by experts for laypeople. It includes titles on mathematics, algebra, numbers, infinity, probability, statistics and logic, among others.

2 **'The unreasonable effectiveness of mathematics in the natural sciences'** is a seminal essay written in 1960 by the physicist Eugene Wigner. It is available at http://www.maths.ed.ac.uk/~aar/papers/wigner.pdf

3 *Mathematics Made Difficult* by Carl E. Linderholm (1971) is a book for those who appreciate the perverse side of mathematics. It consists of mathematical proofs for obvious statements – made as complex as possible.

4 **'100,000 digits of π'.** If you want to look for patterns in the digits of π, the website http://www.geom.uiuc.edu/~huberty/math5337/groupe/digits.html lists the first 100,000.

5 **'The largest known primes – a summary'** details the largest known prime numbers and the search for them. Go to http://primes.utm.edu/largest.html

6 **Wolfram MathWorld** is an extensive online maths resource giving definitions and background for a range of mathematical concepts. Go to http://mathworld.wolfram.com/

7 *New Scientist*'s **website** has an extensive archive of articles and is regularly updated on all themes mathematical and scientific. Go to www.newscientist.com

Glossary

Complex number Any number involving real and imaginary components.

Imaginary number A real number multiplied by the imaginary unit i, the square root of -1.

Integer numbers The natural numbers plus the negative whole numbers, $-1, -2, -3, -4, -5\ldots$ The sets of the natural and the integer numbers are both countably infinite.

Irrational numbers Numbers that cannot be expressed as a fraction of two integers, however larger. Irrational numbers, such as π, Euler's number e and the square root of 2, can never be written out in full: they have an infinite number of decimal places with no repeating pattern in them.

Natural numbers The countable numbers $1, 2, 3, 4, 5\ldots$, generally including 0, too.

Prime numbers The subset of the natural numbers that are greater than 1 and only divisible by 1 and themselves. The set of prime numbers is countably infinite, too.

Rational numbers Numbers that can be expressed as a fraction of two integers, e.g. $1/3, -3/14$.

Real numbers The integer numbers plus all the rational and irrational numbers in between, forming a continuous number line. The set of real numbers forms the continuum infinity, which is larger than countable infinity.

Transcendental numbers The subset of irrational numbers that cannot be made into an integer number by any mathematical operation, e.g. multiplying by itself, raising to a power. π and e are the most famous examples.

Picture credits

All images © *New Scientist* except for the following:

Figure 1.1 Folger Shakespeare Library Digital Image Collection

Figure 2.2 Universal History Archive/Universal Images Group/ REX/Shutterstock

Figure 4.2 John D. & Catherine T. MacArthur Foundation

Figure 5.1 Mint Images/REX/Shutterstock

Figure 7.1 Artem Panteleev/Interpress/TASS

Figure 7.4 AZ Goriely

Figure 8.3 Joel Haddley

Figure 8.4 Courtesy of the Urban Redevelopment Authority, Singapore

Figure 8.5 Elkanah Tisdale/Boston Centinel, 1812

Figure 9.1 F. Schmutzer, restoration by Adam Cuerden/Austrian National Library

Figure 9.2 REX/Shutterstock

Figure 9.4 Jbourjai – Mathematica output, created by author, Public Domain, https://commons.wikimedia.org/w/index. php?curid=5249718

Index